GOLD, GHOSTS & SKIS

Lost Sierra

The legendary days of skiing
in the California mining camps
with 231 photos and maps

❧

William Banks Berry

edited by
Chapman Wentworth

Acknowledgements

My thanks to historians, writers, museums, clubs and friends who have contributed to this history. Special thanks to those who have contributed photographs, which give vivid glimpses of the Snow-shoe Era, 1849 - 1911.

Bill Berry

Photo Credits

All Photographs, unless otherwise identified, are from the author's collection. Reproduction is forbidden without the author's permission.

This book was designed by Susan G. Crowell, who also served as a consulting editor. William P. G. Chapin, copy editor; Janet B. Wentworth, copy consultant.

Published in the United States of America
by the Western America SkiSport Museum,
P.O. Box 729, Soda Springs, CA 95728

Copyrighted 1991, William Banks Berry & Chapman Wentworth

All rights reserved. No part of this book may be reproduced without the author's permission except brief portions used for publicity purposes.

Library of Congress   ISBN

Dedication

This book is dedicated to countless sportsmen and supporters of organized skiing I have known and worked with over my 70 years as a ski enthusiast, supporter, reporter and historian. The following is only a partial list of those, both living and dead, to whom I wish to pay my respects:

First, to professional editor Chapman Wentworth, without whose effort this book would never have been published.

Harold Grinden, former president, National Ski Association (NSA) and historical committee chair, who hoped to write a U.S. Ski History & its relationship to Canada; Roger Langley, former president NSA, honored by National Ski Hall of Fame, Ishpeming & Canadian Ski Museum, Ottowa, Canada. Other NSA Presidents: Al Sigal, William McClure, Byron Nishkian and Red Barth; FWSA Presidents: Paul Smith, Chris Schwarzenbach, Sutter Kunkel, Stan Walton, Richard Goetzman, John Watson, and Richard Croft.

Otto Schneibs, Austrain War II machine-gunner, Dartmouth ski coach, who proposed guidelines which culminated in National Ski Hall of Fame biographicals that qualified its first 150 members.

Enzo Serafini, Buck Erickson, Jakob Vaage, Norway's King Olav V, U.S. Senator John Warner, California Senator Swift Berry; Vicki Nash, Thorwald Rafoos, James Sinnott, Wayne & Sandy Poulsen, Madelyn Walton, Linda Winthers, Tony Wise, Vaughn Webb, Robert Wood, Russ Tiffany, Knut Solvang, William P.Tindale, Vicki & Beatrice Hasher, Lynn Johnson, Hal Codding, Harry Rosenberry, John Robie, Wendell & Inez Robie; Roy & Esther Mikkelsen; Bernard Mergen, Ph.D. J. Stanley Mullin, Ed Tyson, Leroy Rust, Lee Olson, Richard Moulton, John & Heidi Allen; Sally Lewis, Robert Baumrucker, Dorothy Egbert, Mr. & Mrs. Leonard O'Rourke; Elizabeth Merian & family of Onion Valley, Poker Flat, La Porte and Downieville; Fay Lawrence, Ray Leverton, Phillip Earl, Einar Haugen, Hjalmer Berg, Mrs. Jerry Burrelle, Gretchen Besser, Les Bodine, Axel Davidsen, Lynn Douglas, Patricia Armstrong, Rosemary Phillips, John Clair, Bob Parker, Vi & Carson White; Laurance & Ann Maurin, Harold Hirsch & his mentor, and John McCrillis.

Organizations: Norwegian Ski Museum, Norway; Show-shoe Thompson Lodge, Sons of Norway, Yuba City; Douglas County, NV, Historical Society; Nevada Bi-Centennial Commission, Far West Ski Association, Lotta Crabtree Foundation, Boston; Tshopp Family, Nevada Historical Society, Nevada County, CA, Historical Society; Plumas County Museum, Downieville Museum, British Columbia Sports Hall of Fame, Rossland Ski Museum, British Columbia.

Newsmen: Frank Elkins, N.Y. Times & Long Island Free Press; C.J. Lilley, Sacto Union; Eleanor McClatchy, Sacto Bee; Rod Stollery, Tahoe World; Sam Wormington, author, The Ski Race; Henry Pflieger, historian and Stella Riesbeck.

Table of Contents

Chapter 1
setting for a legend — *1*
snow-shoe era in La Parte results in the birth of skiing in California

Chapter 2
eve of an era — *12*
fable of a lake of gold draws thousands to challenge the Lost Sierra winters

Chapter 3
wipsaws for spruce — *26*
pioneer tools hew snow-shoes and lightning dope adds speed to the first snow-shoe racing in the Lost Sierra

Chapter 4
quicksilver charlie — *36*
snow-shoe clubs dominate skiing and races are formalized – California's first ski writer

Chapter 5
stake racing days — *46*
miner's race - mining camps claim "firsts" in snow-shoeing history – racing rivalries draw big crowds and big bets

Chapter 6
the alturas snow-shoe club — *58*
La Porte is the racing wheel of the club – first "world champion" more communities form clubs – new speed records

Chapter 7
the man from silver mountain — *72*
first mountain winter mail routes

Chapter 8
a silver link — *84*
Snow-shoe Thompson's defeat at racing – editorial duels

Chapter 9
dope was king — *96*
dopemaker's recipes set new ski racing records

Chapter 10
volunteer corp — *108*
stories of heroism, broken bones and avalanches

Chapter 11
   the luck of Bob Francis   *116*
      skier's rivalry brings death to the loser

Chapter 12
   the feather   *126*
      a collection of old-timer's legends

Chapter 13
   southern threshold on the yuba   *140*
      men of destiny and new county lines

Chapter 14
   table rock   *148*
      burials in the snow – new racing champions emerge – the best woman snowshoer

Chapter 15
   the ladies   *156*
      civilizing the Lost Sierras with matrimony and courage

Chapter 16
   miners unite in spirit   *166*
      fraternities and churches

Chapter 17
   tracks of the iron horse   *176*
      the first snow trains – challenges of railroad construction

Chapter 18
   a ski center in the sierra   *194*
      the Lost Sierra becomes a destination for sports and recreation

Chapter 19
   the lost sierra revisited   *208*

    Appendices:

       No. 1 .... *i*
       No. 2 .... *iv*
    Lost Sierra Photo Portfolio .... *vi*
    Index .... *xvi*

Introduction

It is said that men on horseback settled the West, but in Bill Berry's "Lost Sierra" man's ingenuity fashioned generations of strange-looking shoes first able to conquer the deep snows, then finally capable of speeds previously unknown to man - helped along by "lightning dope".

Bill Berry's "Lost Sierra" brings in the colorful background of the 19th century in this never-before-told account of the evolution of skiing in the Far West during the "Snow-shoe Era", when even horses wore snow-shoes. The history spans more than a century: from the coming of the '49ers to the development of one of the largest ski complexes in the world.

Berry, a well-known sportswriter and journalist, writes in the lively personal style similar to the fiery frontier editors who reported the "Snow-shoe Era". In researching the book, Berry travelled into the still wild Lost Sierra for his sources. There, among greying timbers he found oldtimers who had settled in cozy mountain cottages for generations. They remembered "back when" - to a time when winters shut down towns for weeks, and skis or "snow-shoes" were 12 feet long, or more. He discovered firsthand accounts of dramatic rescues in the snows, wild west shootouts and racing triumphs.

As he relates in the final chapter of the book, nature in this part of the Sierra Nevada expects every man to hold his own. As an enterprising journalist, Berry himself had to outwit the harsh winter storms in search of this regional history that tells of the true beginnings of downhill ski racing in America.

Berry's personal approach to telling a story makes this book a treasure of unmined stories for anyone interested in ski history or stories of the Far West. His painstaking research through newspaper files from the 19th century and forgotten documents is supplemented by accounts of those who helped build skiing in the Sierra Nevada into a worldwide tourist attraction and competitive racing arena. These accounts evolved from personal friendships with western ski heroes, from ski racers to politicians.

Although skiing has become big business at the snowline, The heart of the Lost Sierra, pocked by ghost towns, slumbers today inviting nature lovers' willingness to travel back roads to explore its hundreds of square miles. In the mountains lie lodes and ledges from which great riches of gravel mines were once derived. On ridges to the south, east and west still stand hundreds of millions of feet of the best sugar pine timber left in California. For 100 years the mines and forests of the Lost Sierra poured out wealth to the world. Mint records show that during a 17-year period the Lost Sierra shipped $55 million in gold bullion to the United States Mint in San Francisco. That was in addition to the millions that went home in the pokes of the Argonauts. Over $93 million came from the Lost Sierra during the Gold Rush era which came with a fury, then left behind only the ghostly echo of the once familiar snow-shoer's greeting, "hey snow-shoer, how's your dope?"

Many among thousands of hopeful miners died as they clambered up the forbidding slopes of the Lost Sierra, a high rugged plateau that lies directly in the path of California's winter storms. The miners were drawn by tales such as that related by a ne'er-do-well named J.R. Stoddard, who told before the threat of hanging of a lake where gold nuggets could be scooped by the handful along the shore. Yet, thanks to the discovery of snow-shoes, many more came to live in the Lost Sierra and made their fortunes. They became familiar names in history or settled in mountainous mining towns with names like Spanish Flat, Rabbit Creek, Rich Bar, Saw Pit, Whiskey Diggings, Onion Valley, Port Wine, Gibsonville, Johnsville, Henness Pass, St. Louis, Poverty Hill and finally La Porte, first called Rabbit Creek House, a town that birthed an 88 mph speed record in racing.

Necessity was the mother of invention. Snow-shoes appeared in 1853 shortly after one of the worst winters in history the year before where miners made epic escapes, such as that made by Hamilton Ward and James Murray, credited with gold discovery in 1850, who fled La Porte, then Rabbit Creek on barrel staves. Acts of heroism were aroused by violent snowstorms and avalanches, while others barely survived harsh conditions and hunger with no supplies in the mining camps.

But pioneer families continued to come the next spring. They entered the Lost Sierra by wagon train via Johnsville, Henness Pass and Yuba River. With them came tools such as throw-irons and whipsaws for hewing snow-shoes from logs. These snow-shoes became the only way to socialize, for physicians to make housecalls or, most spectacularly of all, for mail carriers, receiving only tips or poor pay, to brave relentless storms to bring news of the outside world. Descendants of some of these carriers still continued to use snow-shoes as a mode of travel into the middle 20th century. The mailcarriers in the Lost Sierra became the first true heroes. Life in the Lost Sierra centered on the efforts of men like Zacharia Granville, snow-shoe expressman in 1857, Fenton B. Whiting who went from snow-shoes to dog teams and horses with snow-shoes, and the most famous mailcarrier of all, Snow-shoe Thompson, who carried Virginia City ore samples to Sacramento and may have triggered world attention to the Comstock Lode.

Shortly afterward, new snow-shoe heroes were found as racing fever took hold of the region. Crowds of miners bet their stakes on faster and faster speeds which came about because of secret "dope" recipes applied to the bottom of snow-shoes the way ski wax is today. Heated rivalries even led to infamous killings such as the often told legend of racing heroes Robert Francis and Robert Oliver.

In the second half of the book Bill Berry gives an inside look at the influence trains, cars and other modes of transportation had on snow-shoe travel, the formation of racing clubs; their demise, then revival in the early 20th century by dedicated snow-shoers, and finally the people responsible for building a multimillion dollar ski industry in the Sierra Nevada.

He tells the inside story of how fledgling ski clubs convinced the powerful International Olympic Committee that California had an even richer wealth than the gold mined in the 19th century. These riches found at the snowline were capable of attracting visitors from around the world to the Sierra Nevada mountains.

For many decades Bill Berry, journalist, sportswriter, author and consummate skier, has been on the frontlines (or the snowlines) in his journalistic interpretations of far western history in the 20th century. As a New York Daily News correspondent, it was Berry who through his story of the famous Myron Lake divorce, brought widespread attention to Nevada as the divorce capital of the world.

William Banks Berry was born April 7, 1903 in Potsdam, New York. His newspaper career began at age 13 with Canada's Ottowa Journal. Ottowa Journal's shop foreman taught Berry how to run a Linotype, a skill which got him into the International Typographers Union and assured him a newspaper job wherever he went. Editors quickly spotted his reporting skills. Berry's career has included reporting on world-breaking headlines such Charles A. Lindbergh's 1927 takeoff for Paris covered for the New York Herald Tribune.

In 1928 he moved to Reno, set type for the Nevada State Journal and launched a free-lance news service. He hooked up with INS, UPI, AP Photo and major U.S. dailies. His copy grabbed page one bylines due to his celebrity coverage of Reno as, "The Divorce Capital of the World." James D. Scrugham, Nevada State Journal publisher, took Berry to Truckee's Hilltop Ski Area in December 1928. Soon Berry's ski pieces appeared. His interest in skiing led to the "Lost Sierra" and the subsequent founding of the William B. Berry Western America SkiSport Museum at Boreal Ridge, Donner Pass, on I-80. Berry is a member of the U.S. Ski Hall of Fame. He is Historian Emeritus of the U.S. Ski Association and the Far West ski Association. In 1991, at age 88, he is the longest living active member of the U.S. Ski Association, a charter member of the Reno Ski club, since 1930; a member of the Auburn Ski Club, since 1933, and a charter member of the Far West Ski Association. He has skied since 1910. Berry has won every ski journalism award there is and is an honorary lifetime member of the North American Ski Journalists Association (NASJA).

The original manuscript for this book was written in 1952. Additions followed in 1955 and 1962 as he revived, recreated and reported "Snow-shoe Era," longboard races, as you will see.

Chapman Wentworth

1

The Lost Sierra

*"....Few outsiders ever enter the ski cradle today
(The Lost Sierra) as it is virtually a lost world where life
has not changed since gold was discovered
on Rabbit Creek in 1850."* SPORTS ILLUSTRATED, 1938

Miners gathered at the Snow-shoe Flat, cocktail bar near Johnsville, for this 1897 photo. Their championship racer with a star on his cap is seated in the foreground. W.B.B. Collection.

When late autumn snow flies in the High Sierra, drifts quickly bury field and forest in their deep, silent embrace. The frozen mantle takes charge until late spring. The Lost Sierra rugged terrain transforms into an inland sea of wind-piled drifts hiding weathered relics of historic mining towns and camps. W.B.B.

Scattered across a high, corrugated plateau, surrounded by California's Sierra Nevada, there lies a forgotten land known as the "Lost Sierra", now a crumbling patchwork of once-teeming, turbulent mining camps. This region of the Sierra holds the memories of hardy pioneers who conquered the harsh elements on skis and set racing traditions that remain inspiring today.

Miners, lured by a legendary "Gold Lake", once dug, scraped and sifted over $93 million in gold from The Lost Sierra's streams, diggings, and mines through four colorful California Gold Rush decades of placer and hydraulic mining that began in the early

1

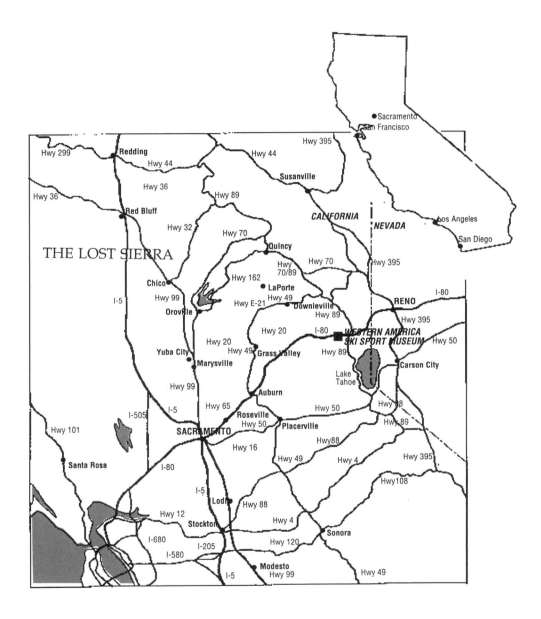

Today the Lost Sierra has become a remote, almost forgotten mountain region, far from the busy main highways that cross the Sierra. It's an outdoor paradise. Natural mountain barriers and snow-choked roads have kept the Lost Sierra separate, a world unto itself, for 100 years. There are no major highways through its remote reaches. It's an enclave, locked within higher elevations of Plumas and Sierra counties. The high plateau is a forested section of California which separates its fertile central and foothill valleys from the Nevada deserts. This virtually forgotten world is bordered on the north by the Feather River's North Fork, split by the Feather's Middle Fork, one of the deepest and most inaccessible canyons in America; guarded to the south by the Yuba's North fork. To the east jut the ramparts of 7,540 foot-high Eureka Peak, the mountain whose image graces the great seal of California. It looms like a seemingly impassable battlement, guarding the eastern approach to the Lost Sierra.

1850's. In fact, it was in the heart of the Lost Sierra near La Porte that the largest nugget in the U.S. was found. Weighing slightly more than 313 pounds, the nugget was then valued at $84,302.

Since the region stretches directly across the path of the Pacific Ocean's savage winter storms the "Lost Sierra" challenged man's capacity to survive. In winter raging blizzards howl across mountain-studded forests. On this high land-locked plateau, stretching from La Porte to the Feather River on the north and the Yuba River on the south, Forty-niners either coped with the elements or perished. Man's ingenuity and persistence ultimately won with the birth of the Snow-shoe Era which helped create pioneer settlements during the mining decades.

La Porte was once the thriving hub of this gold country. Other ghost towns from the gold era on the plateau were Gibsonville, Howland Flat, Poker Flat, St. Louis, Poverty Hill, Scales, Port Wine, Whiskey, Diggings, Hepsidam, Nelson Point, Onion Valley and a score of other picturesquely-named diggings.

In 1883, when hydraulic mining was outlawed by the state of California the gold mining industry collapsed and bustling communities began to die. Meccas like La Porte, once boasting some 3,000 people, dropped to no more than 25 year-around residents. Others, such as Onion Valley, a once vigorous mining community, disappeared.

This is one reason why the region is referred to as "The Lost Sierra" by historians who have studied the rich culture that thrived from the late 1850's to the late 1880's.

Out of this once fabled gold producing region has emerged a vigorous legend of snowbound life as it was experienced during and immediately following the Gold Rush. Folklore again and again tells the story of a strange race of snow-shoers, thousands strong, which once inhabited the high Sierra. Now we know they were the Forty-niners in winter.

What the miners had then called "snow-shoes" are today called skis. Around La Porte they still call them that. Use of longboard snow-shoes first became widespread in what we now call "The Lost Sierra." This high plateau between the Feather and the Yuba Rivers includes parts of Plumas and Sierra counties. It is aptly called the cradle of skiing in California.

The longboard travelling and racing snow-shoes, described by the pioneers as made of slim clear-grained spruce billets, were actually made from Douglas fir *(Pseudotsuga douglasii)* since there is no commercial spruce in the Sierra Nevada. Use of the snow-shoe evolved from the need for winter survival. They also served as social convenience, offering a means for short and long distance winter travel during one of American life's most romantic epochs.

Unknown to the outside world, snow-shoers appeared in the wake of the '49 Gold Rush and continued their heyday during the four decades that followed. The Snow-shoe Era dates from the appearance of snow-shoes in the 1850's, until the miners' famous last race sponsored by former snow-shoe champion Frank Steward at La Porte in 1911.

Snow-shoes, sometimes called Norwegian snow-shoes, came to the mines with the Argonauts: the early California gold seekers. Their strange snow-shoe tracks were first seen in the Lost Sierra. Soon they were cutting snows the length and breadth of California's mountain ranges. Use of snow-shoes spread northward to the Cascades. They were soon picked up in Oregon and Washington. Eastward too, they became popular in

Mail sleigh breaks trail through the forest near La Porte.

one prodigious leap, appearing in the far-flung mining camps of Utah, Colorado, Idaho and Montana.

In Yellowstone National Park, they were used by U.S. Army troops. Soldiers served as the first park rangers. These snow-shoes were the ski kind rather than the web-type, also called snow-shoes used by Indians.

Longboard snow-shoes played an important role in the early conquest of the American West. Because of the extraordinarily deep snow in the high country of Plumas and Sierra counties, the use of longboard snow-shoes was interwoven with the pioneers' life, work and recreation.

It was cradle-to-the-grave snow-shoe life. Doctors traveled miles on snow-shoes to deliver babies who in infanthood learned to ride snow-shoes almost as soon as they could walk. Children traveled to school on snow-shoes; adolescent romance blossomed at snow-shoe races and dances that followed. Newlyweds made honeymoon trips on snow-shoes. Men and women found recreation in frequent snow-shoe tournaments. Men snow-shoed to work and wore them to deliver mail and supplies.

Early-day and even present-day writers, who painted glowing word pictures of California's mining camps, avoided the snow-gripped winter rigors of the Snow-shoe Era.

Their writing has been limited to tales of life in the mines observed during the warm spring, summer and autumn months.

There appears to be a plausible explanation for the omission of what transpired during the heavy snow of

Snow piles around the Thistle mine shaft near Gibsonville. With mines closed in winter, miners hit the slopes on their snow-shoes.

California's Sierra winters. There was a social schism locked into the '50's and '60's culture. Society drew a dividing line cut between laborers in the mines and San Francisco's nouveau riche. During winter's reign, literati of those times were wont to foregather in softer, more luxuriant surroundings during what was popularly called "the rainy season."

No such way of life as the miners had developed around their new-fangled snow-shoes was worthy of mention, even if known about. The writers knew that North American Indians had been using "snowshoes" for hundreds of years. To those writers who had long ago heard all about the Indians' snowshoes, there was nothing new in that!

What mattered it that a race of people, thousands strong, inhabited the mining camps during the winter months? Who cared that they had struck it rich, in a new way, having created a novel means of enjoying a "rainy season" social life?

Affluent mine owners, well-to-do traders and mining stock speculators, or at least a majority of them, remained dry, warm and comfortable in their San Francisco habitat.

Wintering in the high Sierra was left for day laborers. The mountains harbored hundreds of lowly miners who had failed to grub enough gold from their diggings to afford to leave. Others who stayed must have been regarded by the media as just plain damned fools! Nevertheless, as it blossomed on the frontier, the new way of life spread far and wide. Snow-shoe mail carriers and racers became heroic figures in their own bailiwicks. Most notable was John A. (Snow-shoe) Thompson of Alpine County, the famous winter mail carrier from Silver Mountain. Thompson and others basked in the press which slowly spread the glory of their feats.

They burgeoned out of the Lost Sierra of Plumas and Sierra counties. Swarming across the Feather River and Yuba River Canyons, they spread far and wide. From Yosemite they traveled across the Tioga, from the San Joaquin valley, across the Sonora, Ebbetts and Kit Carson passes. They snow-shoed from Hangtown east to the Utah Territory and north over other mountain passes. Soon these longboard riders had conquered the Sierra in routine crossings. A network of blazed trails led through pine forests.

New-fallen snow found miners and mountaineers cruising along on their snow-shoes to visit neighbors. Or they headed towards racing and recreation centers in the roaring camps of Plumas and Sierra. Deep snow was no longer a peril.

This virtually forgotten world is bordered on the north by the Feather River's North Fork,

Historic picture of snow tunnel March,1906, at Harris' store, La Porte.

split by the Feather's Middle Fork, one of the deepest and most inaccessible canyons in America, guarded to the south by the Yuba's North Fork. To the east jut the ramparts of 7,447-foot-high Eureka Peak. The mountain, which image graces the great seal of California, looms like a seemingly impassable battlement, guarding the eastern approach to the Lost Sierra.

There is scarcely a drop of water or a flake of snow that falls within the Lost Sierra which does not find its way into the tributaries of the Feather and the Yuba. Moisture falls in the region chiefly as snow. The Lost Sierra frequently finds itself piled under a snowpack as much as 15 feet during winter, storing precious moisture for California homes and farms.

Surging in from the North Pacific each autumn roars one storm after another. Each brings heavy rain to the Sacramento Valley. Blizzards paint lofty summits of Plumas and Sierra with gleaming white. Ten-foot snow depths are commonplace in the Lost Sierra. Twenty-foot depths are not unusual. Even drifts from 30 to 50 feet deep have been recorded. During most years in early fall, roads in the foothill valleys were barely passable, because of mud. At higher elevations, mountain roads were even more solidly blockaded by drifting snow, even further isolating the pioneers from civilization.

Across the ridges, separating early-day mining camps, the only means of communication was the delivery of mail and express by carriers on snowshoes. The snow-blocked routes of travel created problems for the pioneers. It fell most cruelly on them during the early years. Those caught unprepared by freezing blizzards, buffeted as they trudged through mountain passes, endured great suffering and often

Even horses wore snow-shoes during The Snow-shoe Era in the Lost Sierra. Above, March 5, 1906 photo, shows snow-shoe equipped team arriving at La Porte with freight sleigh ladened with mail and provisions.

death. As the story of the ill-fated Donner Party tells us, many perished far from help or friends.

Others, almost within reach of shelter, collapsed from fatigue and cold, or were swept to their deaths in roaring avalanches.

Two years after the discovery of gold in California, Argonauts invaded the Lost Sierra. Gold seekers pouring through the Golden Gate included Yankees, Irish, Scots and English, Scandinavian, Oriental, Kanaka, gentiles, Jews and men of many faiths. All swarmed, hellbent in 1849, toward the foothill diggings.

It was then that the "Lost Sierra Rainbow," a fable in the form of a mythical golden-sanded lake, beckoned. That spring of 1850 thousands of miners stormed the snow-covered Lost Sierra in pursuit of this dream. None ever found the goldstrewn shores.

However, as soon as the mythical character of the stampede's objective was proclaimed, disappointed gold seekers found their way to truly profitable diggings. They prospected throughout the mountain canyons. Trees were felled and clearings cut. Tents and canvas-covered shacks rose and later gave way to log cabins. Soon homes appeared as towns and trading centers became established.

Miners abandoned their claims, heading for the

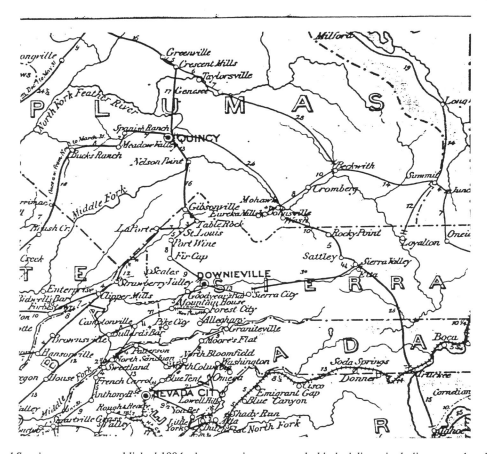

U.S. Postal Service route map, published 1884, shows carrier routes as bold, dark lines, including cross-hatched railroads, with mileages noted in between. The light line from Downieville to St. Louis, then dark to La Porte and Table Rock, is a tough route, usually interrupted in winter. The main route through the "Lost Sierra," was from Marysville, through La Porte, to Quincy. Nevada Historical Society photo.

lower valley safety during the winters of 1850-53.

During the summer of 1853 snow-shoes were introduced and the Snow-shoe Era began. The event went unnoticed by the media. Emigrants had poured into the valleys surrounding the Lost Sierra. Soon these pioneers transformed it from a wilderness into an abode for man. Prospectors swarmed to the high country. They soon discovered snow-shoes as the only creative way to travel in winter. The emigrants took to agriculture and the miners to snow-shoeing. With snow-shoes handy, miners no longer fled to the central valleys at the approach of winter.

What tales adventure-seeking authors such as Bret Harte and Mark Twain could have written had those romantic writers of early-day California mining days seen this high country in winter. What vivid stories they would have spun had they observed the strange customs and traditions of its first inhabitants.

Happily, the local color still remains, even though snow-shoeing gold seekers are gone and a majority of old timers rest in peace underneath the snow. A handful remain where once thousands mined the gold channels and dared the snows, but they carry on in hardy pioneer fashion.

topography

The topography of the Lost Sierra presents a continuous succession of lofty ridges and deep canyons. Ridges rise to dizzying heights and hundreds of the canyons sink into the bewildering depths. Forests of

Miners poised at the top of race course, ready to push off when starter, at right with hand raised, hits the gong. Famed La Porte "dope" maker, Frank Steward, far right, has waxed their longboards. Photo circa 1905, Plumas County Museum collection.

red fir, balsam fir, cedar, sugar and yellow pine abound. They once provided shelter for the early-day mining camps situated at elevations ranging from three to six thousand feet.

Spring's melting snows feed numerous crystal torrents on every side. Tumbling waters blessed the gold miners' work in early day diggings. The torrent supplied water for hydraulic mining which followed the era of placers, rockers and long tom operations. In addition to these streams, there are many small, lovely lakes scattered throughout the region.

The isolated mountain peaks of the Lost Sierra include Table Rock, Pilot Peak, Eureka Peak and the Sierra Buttes. Table Rock, sporting a famous snowshoe racing course of early days, reaches an altitude of 7,050 feet. Pilot Peak reaches 7,605 feet, Eureka Peak juts 7,540 feet. The miners found plenty of challenging ski terrain. Sierra Buttes, one of California's most famous landmarks, etched the sky at 8,950 feet.

On the slopes of Pilot Peak, an isolated volcanic knob, the very first organized ski competition ever known in America took place in the mid-1850's. The racers came from Onion Valley, 1,216 feet below the peak's summit. Nowhere in the Sierra is there a more stupendous example of denuded slopes than occurs in the region north and northeast of Pilot Peak. At Nelson Point sheer walls of rock tower some 3,650 feet above the Middle Fork of the Feather River, Nelson Point where gold was discovered on the middle fork, once held sway as a famous snow-shoeing center. In the early days it straddled the main route for the snowshoe mail carrier and packers. They passed it as they slid with abandon down the slopes.

View of La Porte's main street in 1858, once boasting some 3,000 population. (Reproduced from an old engraving.)

Deep snow had covered homes to the eaves in Johnsville, Jan 20, 1913. Eureka Peak is seen in the background. The famed peak, immortalized on the Great Seal of California, is the center piece of Plumas-Eureka State Park.. Photo: Lynn Douglas collection.

the birth of snow-shoes

Here it was that gold miners found that snow-shoeing could be fun as well as a way to get work on the silvery snow.

Whether what the pioneers called snow-shoes, but actually were skis, were introduced by Scandinavian sailors who had deserted their ships in San Francisco Bay to join the first mad rush to the gold diggings in 1849, or whether they were a result of Yankee ingenuity, remains to this day moot in the Lost Sierra. The old-timers declare that the very first snow-shoes were staves taken from flour barrels and used by Hamilton Ward who had "struck it rich" in 1850 in the gravel banks along Rabbit Creek, later called La Porte.

One "Lost Sierra" yarn tells us Hamilton Ward and James Murray used barrel staves to escape from the mountains when early-season storms came. Some old residents have insisted their ancestors then improved upon that means of locomotion. The first wooden snow-shoes are said to have been very crude affairs, providing a man with not much better than walking support across the snow. Then during the winter of 1853-54, came the introduction of "Norway snow-shoes" or "Norway skates." These again were improved upon to fit Sierra snow conditions. From these meager beginnings the region's pioneers developed cross-country skis 8-10 feet long and also lightning fast racing models,10-12 feet long, called "traveling and racing snow-shoes."

The transition of the snow-shoe from utility use to sport was rapid. Perhaps the free mountain air had something to do with it. It may have been the moral influence, so badly needed on the borders of civilization during gold mining days. But in any event, with snow-shoes man moved faster than he ever had.

The setting for the legend in the Lost Sierra of Plumas and Sierra counties can now be told. That cradle of tall peaks and deep valleys is the birthplace of High Speed, Downhill Ski Racing. A sport that created a form of organized companionship which now lures millions to the world's snow-covered slopes.

In the Lost Sierra, America's first ski clubs were formed for the specific purpose of staging downhill competition. It was in this historic mountain locale that skis were first manufactured on a commercial basis in America. It is where "dope", now called ski wax, was developed and where the world's very first organized ski races were held.

The Lost Sierra is the place where skis and skiing evolved to a point comparable to any achieved in this modern day. There, scattered through the towering mountains, live the last of a fabled people, the snow-shoers of the High Sierra.

References:

La Porte Scrapbook, by Helen Weaver Gould, 1972
Hutchings California Magazine, December, 1860.
Illustrated History of Plumas, Lassen and Sierra counties, by Farris & Smith, 1882.
California's Pioneer Mountaineer of Rabbit Creek, by Albert Dressler, 1930.

2

Eve of An Era

> *"During the long winter months the accounts of Stoddard's 'strike' spread through the foothill mining camps the entire length of the Sierra. So strange was the tale, that many believed him crazy, but Stoddard's samples provided a convincing argument."*
> —Farris & Smith, 1882, "Illustrated History of Plumas, Lassen & Sierra Counties from 1813 to 1850."

Perhaps it is paradoxical that the man responsible for triggering the "Lost Sierra" gold rush never wore a pair of snow-shoes.

The legendary figure, a man known simply as J.R. Stoddard, played a role that led directly to opening up and settling the Lost Sierra.

He missed the thrill of skiing, but certainly had excitement enough when a throng of angry miners measured his neck for a noose.

It may appear ironic to us today, as we read the Stoddard gold rush fable in Farris & Smith's vivid history of Plumas, Lassen and Sierra counties, that several of those who had set themselves up as "Judge Lynch" to complete the California-style necktie party for Stoddard, may have met their own deaths in those selfsame mountains simply for the lack of snow-shoes.

Nevertheless, it was out of Stoddard's adventure, followed by miners' wintertime suffering and death, that the Snow-shoe Era evolved.

Snow-shoeing came to life as a metamorphosis, started by the gold fever's influx of miners. Then, inevitably, as the miners learned to cope with winter, it blossomed with the snow-shoers' excitement as they plunged at lightning speed down the Lost Sierra mountainsides.

The Lost Sierra was then, as it is today, relatively unknown.

A large overland emigration of gold seekers had passed westward across the Sierra Nevada during the summer and fall of 1849. They were unaware of the riches beneath their feet as they struggled across the Lost Sierra region. Some emigrants had been diverted from the Carson route and were induced to follow Lassen's cut-off, or, as it was sometimes called, "Lassen's Horn" route, sarcastically classing it with the perilous trip around Cape Horn.

Of the Argonauts who had actually sailed around the Horn, none had ventured into what is now called The Lost Sierra.

Further, none of the hundreds of overland emigrants making the mountain crossing stayed behind to prospect the streams in that high Sierra country. These latter emigrants' objective was the Valley of the Sacramento. They were determined to get there.

With the mountains behind them, these new emigrants passed camp after camp of miners who were digging for gold in the western slope foothill country.

A few weeks later the truth dawned. These newcomers would retrace their steps. They came back to the diggings, after replenishing their supplies at Sutter's Fort and other trading posts.

For that reason the Lost Sierra welcomed none of the many gold seekers who made the western crossing through it in 1849.

J.R. Stoddard was among these pioneers. He was to become famous as the "hero" of an adventure so shrouded in mystery that it remains a permanent fixture in the region's folklore down to the present day.

Stoddard in the fall of 1849 stumbled into one of the uppermost gold diggings on the Yuba's North Fork with a tale that created what has since become known as the "Gold Lake excitement."

He told a story of having found a small lake, the shores and bottom of which were strewn with chunks of pure gold. He handed out samples he brought with him to prove it. Stoddard told the miners that he had made the discovery after becoming lost from an emigrant train. He had picked up all the gold he could carry. Unfortunately for him, he was pursued by Indians and forced to abandon much of the treasure in his dash for safety.

Discoverer Stoddard was all for organizing an immediate company to return to the lake of gold, but winter had set in. Snow was beginning to pile up in the mountains and none of the miners wanted any part of it.

Snow-shoes had not yet been introduced. It has been said there was a Scandinavian or two among the miners,

California miners in the early 1850's washing out gold in a "rocker," such as was used at Rabbit Creek, later renamed La Porte, capital of the "Lost Sierra." Phil B. Bekeart Collection.

who may have had glimmerings of such an idea along those lines even though not a single American miner could be found who dared trust himself to the winter cold and snow of the upper regions. However, there were many who had faith in Stoddard's tale. It was agreed that a determined effort to seek out the lake and rake in the gold would be made come spring.

During the long winter months the accounts of Stoddard's "strike" spread through the foothill mining camps the entire length of the Sierra. So strange was the tale that many believed him crazy. But hundreds of others placed reliance in the reports. Stoddard's' "samples" provided a convincing argument. At that time many miners had a theory that the "source of gold" lay high up in the mountains. They had noticed that the gold became coarser as they ascended the streams. What was more natural than to assume that there was some place up in the high Sierra where it all came from, and that Stoddard had found the true mother lode.

That spring a select company of 25 men was organized to accompany Stoddard in search of the fabled body of water. About June 1, 1850, the company marched up into the mountains. Following right behind was at least another thousand men, all determined to keep Stoddard in sight, until the end of the world if necessary.

Stoddard led his party to the divide between the North

A meeting of immigrant trains on the plains bound for California

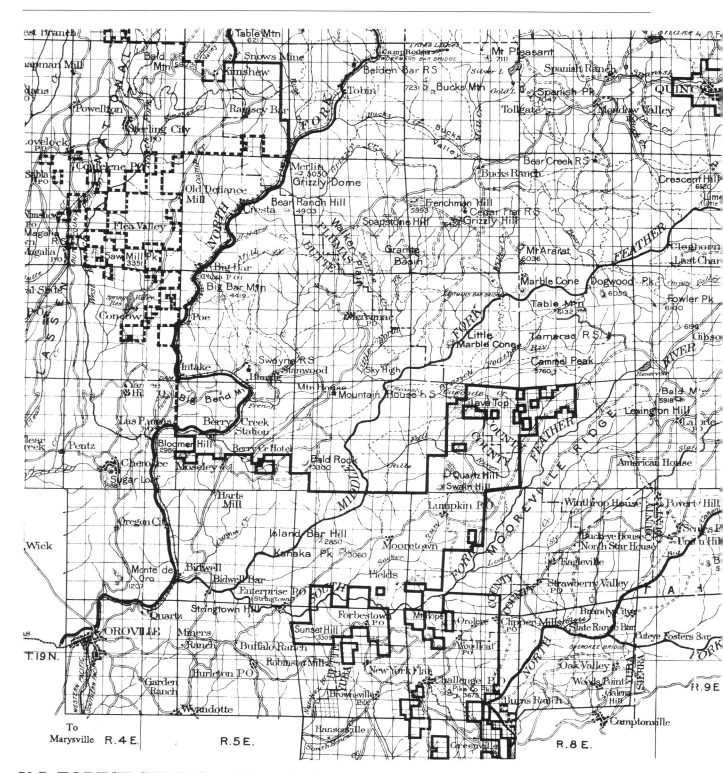

U.S. FOREST SERVICE MAP - 1922

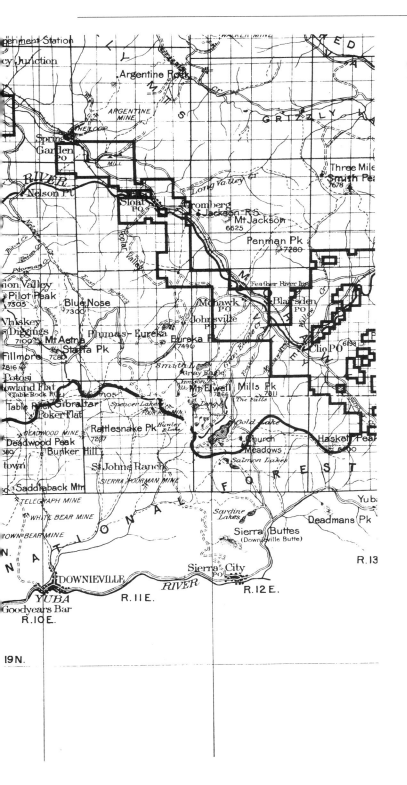

MAP LEGEND

1. Escape from Rich Bar - Some 70 miners escaped death in five-day battle with snow. December 28, 1852. See page 21.

2. Escape of Miner Bain Party - Leaving Soda Bar on New Year's Eve, 1852. Four survivors reached Bidwell's Bar. See page 22.

3. Gold Lake; fact or fiction? - No one ever found J.R. Stoddard's real "Gold Lake." See page 19.

4. La Porte and Lexington Hill: Originally Rabbit Creek. La Porte was the hub of the "Lost Sierra," hosting snow-shoe races at nearby Lexington Hill.

5. Onion Valley: nearby Poorman's Creek, Last Chance Valley, and Saw Pit Flat: became the center of early snow-shoe racing.

6. Gibsonville: One of the major towns in the "Lost Sierra," home of many "Snow-shoe Era" pioneers.

7. Quincy: Located on the East Branch of the North Fork of the Feather River. It was the northern gateway to the Lost Sierra.

8. Downieville: A prosperous mining town, located on the North Fork of the Yuba River. It was the southern gateway to the Lost Sierra.

9. Poker Flat, Howland Flat, Table Rock: close together, between Downieville and La Porte, produced many famous snow-shoe racers.

10. Whiskey Diggings: Prosperous mining camp near La Porte that played a prominent role in the Snow-shoe Era.

11. Poverty Hill: It was here that Dr. Edmund Bryant, first husband of Comstock Queen Marie Mackay, died in poverty. His health was devastated by drugs and drink.

12. Nelson Point: High ridge above Feather River Middle Fork that had to be climbed enroute from Quincy to La Porte.

13. Johnsville: Early mining town on the east slope of the Sierra where folklore says gold was

Continued on page 17

found before the 49ers came.

14. Sierra City: Scene of many of Nevada County's terrible avalanches and a southern gateway to the Lost Sierra.

15. Sierra Buttes: Nevada County landmark, just north of Sierra City.

16. American House: Main stop on the stage coach route from Marysville to La Porte.

17. Meadow Valley: Popular mountain resort near Quincy.

18. Scales: Pioneer mining camp near La Porte that produced a remarkable number of top snow-shoe racers.

19. Bidwell's Bar: Where the Feather River's north and south forks joined, a jumping off point to the Lost Sierra.

20. Strawberry Valley: A pioneer town below the winter snow line on the stage route from Marysville to La Porte.

21. Spanish Ranch: A pioneer trading post where miners working on the Feather River's north and middle forks could re-supply.

22. Buck's Ranch: A trading post on the stage route from Bidwell's Bar to Spanish Ranch and Quincy.

23. Peavine: A popular community between the Feather River's north and middle forks, between Buck's Ranch and Bidwell's Bar.

24. Marysville: Jumping off point for stage and mail routes to the "Lost Sierra."

Yuba and the Feather's Middle Fork, from which point he evinced a firmness of direction that took the company straight to the headwaters. Right behind Stoddard's party streamed the shadowy cloud of camp followers. Stoddard's seemingly purposeful directions soon turned to aimless wandering. It now became evident that he was incapable of leading the way to the wonderful lake. It was a vain search in which all were enduring great hardship.

At what has since become known as Last Chance Valley, on the USFS map near Onion Valley, Stoddard's deluded associates halted for a consultation. Some expressed a belief he was crazy; others concluded he had never visited the supposed lake, but had heard the story from other lips and then represented the adventure as happening to himself in the hopes of forming a searching party; others believed his story true, but believed his sense of direction was poor.

The company was badly disorganized. Stoddard's own select group now had been joined by the thousand others. Many of their pack animals had perished in deep snow while crossing high ridges or had been dashed to their death upon the rocks of precipitous canyons. The original party was overwhelmed by the camp followers. Soon there was open rebellion.

Stoddard can be pictured quaking in his shoes as a miner's court of mountain justice was convened. A decision was reached to hang the author of their woes at once.

However, the sentence was suspended for one day at the solicitation of those few who still believed the lake existed. Stoddard was informed that he had been given a "last chance" to locate the lake or be strung to the limb of some convenient tree and left for the birds to roost upon.

Crazy or not, Stoddard had enough method in his madness to slip quietly away during the night and retrace his steps to the mines below. The "Gold Lakers" were left behind to follow their own inclinations.

This actually was but a small introduction to the excitement that quickly gripped the region. The news that Stoddard and his party, followed by the crowd of miners, had left in search of the gold-strewn lake spread like wildfire through all the California mining camps. The floating population imbibed the fever and soon other thousands were scaling the heights with few provisions and only dust with which to buy. It was a veritable stampede. Prices of horses, mules and oxen soared. Merchants accompanied the eager throngs with loads of provision, which they sold at exorbitant prices They even slaughtered the cattle that drew the loads, selling the meat to the hungry hordes.

The excitement gripped the miners for about seven weeks, then resolved itself into the regular movements of eager gold seekers shifting from old to new diggings. When the original party discovered that their intended victim had fled they reluctantly abandoned the search. They started back, prospecting along the way.

The cloud of followers did likewise, even though the news that Stoddard was a fraud quickly travelled from camp to camp, describing the Gold Lake story as a myth.

They swarmed like ants across the valleys and ridges of the Lost Sierra.

Even before the search for Gold Lake was abandoned by the Stoddard party and its clusters of followers, considerable prospecting had already started among those who came a few days later than the first "Gold Lake invaders". As a result miners reported "strikes" at Nelson Point, Poorman's Creek Spanish Flat, Rabbit Creek and other places.,

Soon the Lost Sierra was swarming with disappointed Gold Lakers. In many cases, in which the first workers had measured off generous-sized claims, the newcomers called meetings, made law reducing claim sizes and proceeded to stake out their own locations. Even this failed to give claims to all. Hundreds who did not secure ground for a mining claim were forced to try their luck elsewhere. In this way the upper forks of the Yuba and Feather and their tributaries were occupied by several thousand miners during the summer and fall of 1850.

It is interesting to note that on the 1922 map of Plumas National Forest, in the heartland of the Lost Sierra, there

A typical gambling scene in old El Dorado Saloon, one of the most famous San Francisco gambling houses of its day.

lies a "Gold Lake" about 10 miles north of Sierra City. It's on a tributary to the Feather River's Middle Fork.

The fear of being caught by winter in the mountains as yet had not been dispelled. So with the first snowfalls a majority of the miners prepared to quit the hills. Soon there was scarcely a sign of animation in the camps where the smoke of hundreds of campfires had scented the clear mountain air. The rattle of gold rockers, once echoing from canyon walls, was stilled.

Many retreated west, down to Bidwell's Bar on the Feather's Middle Fork., Others descended to diggings alongside the lower Yuba. Only a better provisioned few elected to brave the snows in log cabins they had erected.

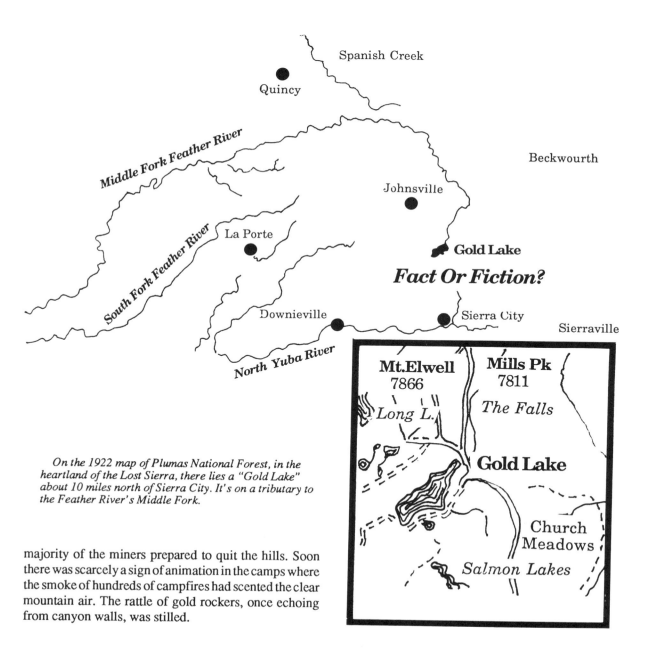

Fact Or Fiction?

On the 1922 map of Plumas National Forest, in the heartland of the Lost Sierra, there lies a "Gold Lake" about 10 miles north of Sierra City. It's on a tributary to the Feather River's Middle Fork.

The winter of 1850-51 found the mines of the snow country practically deserted.

Then came the spring of 1851.

Miners thronged back to the high country diggings, crowding the streams of the Lost Sierra. They spread in all directions as they made new discoveries. Thousands of claims were staked; flumes and wing dams were built to divert water to placer operations; solid cabins were constructed and in many other ways the miners gave indication of their intent to remain year around, at least as long as the gold held out.

Further tokens of permanent occupation came as others took up land claims upon which to establish trading posts.

Several sawmills sprouted and began operations.

Large stocks of provisions were laid in as the winter of 1851-52 approached. Although many returned to the valleys, a majority of the miners, those who planned to work in the diggings the next season, remained in the region. Work was suspended with the coming of snow, but it was a mild winter. For the most part, storms did not greatly inconvenience those who sat it out.

Another influx of newcomers flooded the diggings during the spring and summer of 1852. Many emigrants entered the Lost Sierra from the east, through Beckwourth Pass, where the celebrated mulatto frontiersman, Jim Beckwourth, had built a trading post. The emigrant trains stocked up at Beckwourth's station in preparation for the Sierra crossing, many making it over a trail into Lost Sierra which led past Eureka Peak, which guarded its eastern entrance.

It was during the summer of 1852 that the gentler sex began to appear in considerable numbers. Women and children's presence combined to make the first changes in the rough and ready mining scene. The campfires now began to give way to the domestic hearth.

There was, however, to be experienced, a memorable winter of hardship in the mountains.

the snows of 1852-53

No pioneer of the snows of 1852-53 ever forgot its raw scenes of suffering, death and destruction.

During the summer months the miners' labors had been on the unprofitable side; the cream had been skimmed from the early discoveries. The big cleanups from rockers set up beside the newer gold channels had failed to come up to expectations.

Cautioned by many miners bemoaning their bad luck, merchants failed to lay in large stocks of goods and provisions. Gold dust was to be scarce that winter, but not snow.

The first white mantle settled down in November. The miners believed it was temporary. But snowdrifts were heavy enough to block the mountain trails to all travel. Fortunately the few women and children occupying the higher camps had earlier been sent to lower elevations.

The miners who remained were confident they could stick things out at least until late December. By then, it was confidently predicted, new supplies could be brought in over the mountains.

Then storm followed storm in unprecedented waves. It soon became apparent that all trails had become sealed to pack trains for many weeks to come.

At that time snow-shoes, both the Indian web type and Norwegian ski design, were practically unknown. Travelers had to flounder through the snow as best they could.

A great rush surged from the mining camps as men fought to reach sources of supplies. Many camps were soon deserted, while others were left with only a handful of men. Those remaining behind were only enabled to do so by purchasing supplies abandoned by those who were clambering out via the snowy trails.

Miners from the Middle Fork of the Feather and its tributaries spent the winter in Onion and Strawberry valleys, to which points goods were still arriving. Others scattered themselves among the camps alongside the Yuba River. Those from the North Fork and East Branch escaped to points on the Feather below Bidwell's Bar. Many of these escaped before the dangers of travel became too great. Travelers who started later endured great hardships. Some did not survive.

During that memorable 1852-53 winter, throughout the week between Christmas and New Year's, the mountains were buffeted by a wild series of storms. During that week, and the days immediately following, great suffering and death in the snows befell many miners who had stayed behind. Men who had families in the valleys struggled bravely to go for food they required. Pack trains to Nelson Point on the Feather's Middle Fork came through by use of almost superhuman effort, but even these did not supply enough provisions for the scattered groups of miners who by then suffered near starvation conditions.

On one trip through the mountains a train of mules refused to budge across a snow-covered ridge. The drivers blindfolded the animals, led the train across, then pushed the mules over the canyon side. The mule rolled, packs and all, down to a snowbound camp below.

An epic escape from Rich Bar on the Feather's East Branch began the morning of Dec. 28, 1852. The miners there concluded they must make a bold dash across the barrier of snow or else they would perish from starvation. The snow lay four feet deep in the river canyon. On the mountainside to be crossed, snow lay about 30 feet deep. The drifted snow, four or five feet on top, had newly fallen. When the decision was reached there was but a week's provisions remaining and those were sold to seven men who decided to remain behind. The others, each packing a scant ration of cooked food wrapped in a blanket, started the perilous trek.

There were over 70 miners; American, French, Mexican, Kanaka and Chinese. All began to ascend the mountain. The men took turns plunging forward into the snow to beat down a path. In this way a man would work at the head of the column for a few yards, then step aside to fall in at the rear as the lead rotated. Their first goal was to reach an isolated small cabin six miles distant on the mountain ridge. It was a foot-by-foot struggle. Avalanches roared down. High winds drifted light snow to almost blind the climbers. The elements were too much for some members of the party. Several were forced to abandoned the unequal struggle. They perished and their bodies were left in the snow along the way.

Such a motley crowd never before had lodged in so contracted quarters as jammed the small cabin that nightfall. The snow was over 15 feet deep and the place had only been found with great difficulty. The miners tore up

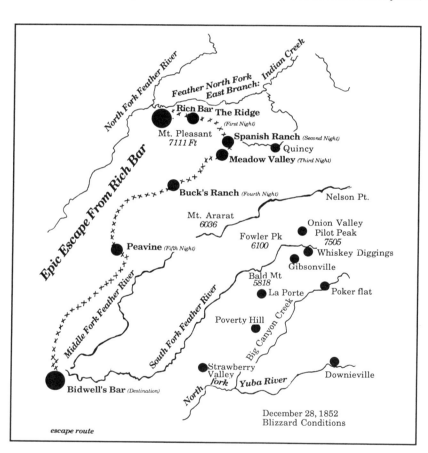

An epic escape from Rich Bar on the Feather's East Branch began the morning of Dec. 28, 1852. The miners there concluded they must make a bold dash across the barrier of snow or else they would perish from starvation. The snow lay four feet deep in the river canyon. On the mountainside to be crossed, snow lay about 30 feet deep. The drifted snow, four or five feet on top, had newly fallen. When the decision was reached there was but a week's provisions remaining and those were sold to seven men who decided to remain behind. The others, each packing a scant ration of cooked food wrapped in a blanket, started the perilous trek.

the plank flooring to provide a fire with which to draw some warmth into their bodies before rolling up in wet blankets in search of the "sweet restorer". Early next morning they renewed their struggle against the snow. Another terrific storm had swept in. Six miles of wallowing brought the exhausted party to Spanish Ranch, where they procured food. Here they found a great crowd of miners who had trekked from other points. The proprietor worried for fear they would eat him out of all provisions.

Ranch, eight miles distant. There they were fed and housed. During the night the snowfall turned into rain. Freezing cold followed, forming a hard crust on the deep snow. Trudging the next 16 miles to a place called Peavine was comparatively easy. Another day's travel brought the company to safety at Bidwell's Bar.

The death of a man known only as Bain was another tragedy during the terrible 1852-53 exodus. Bain had three companions, M. Madden, Thomas Schooly and Mordicain Dunlap. All but the ill-fated Bain were later to

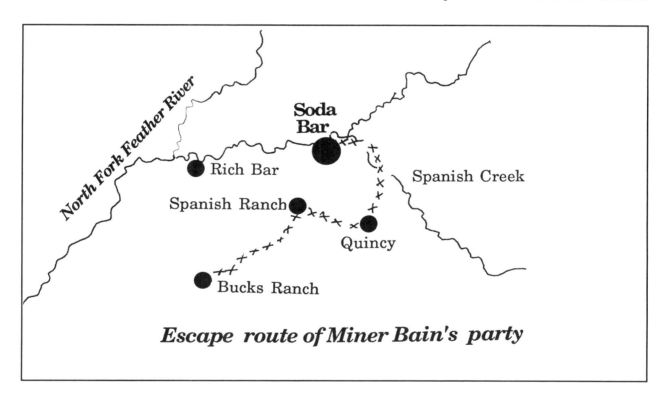

Escape route of Miner Bain's party

The Miner Bain party escaped from Soda Bar to Bidwell's Bar. They started out following the route shown above. From then on the route from the Spanish Ranch was the same as that travelled by the Rich Bar party. Many goldseekers from camps alongside the Yuba River from the North Fork and East Branch escaped to points on the Feather below Bidwell's Bar. Many of these escaped before the dangers of travel became too great. Unfortunate travelers who started later endured great hardships. Some did not survive.

All hands tramped another two miles to Meadow Valley, where the owner of a trading post kept them overnight. He then demanded they move on towards Buck's

become noted snow-shoers. They left Soda Bar on New Year's Eve. They struggled for two days through snowstorms and across raging mountain creeks. At one time they had been forced to ford an icy torrent waist deep in water. Exhausted, they reached Buck's Ranch.

On Jan 2., they began the 16-mile trek through the snows to Peavine. They started out against the advice of older hands.

Bain was the only member of the company who knew the trail. They had only traveled a few miles when they were enveloped in falling snow. The wind blew a gale, biting through to their marrows. None had a full set of clothing. Bain was wearing only boots, pants, and a woolen shirt. The latter flapped buttonless, exposing his chest.

Great difficulty was encountered keeping to the trail, and it was decided to walk four abreast so that the judgment of all could be used. In this manner they struggled along in waist deep snow and reached the ruins of what had once been a canvas shack. Here they dug about with their hands in the vain hope of finding something with which to resurrect the shelter.

A situation to daunt even the hardiest young pioneer now faced the quartet who were cold, wet, exhausted, poorly clad, the darkness preparing to enshroud them, miles away from food, warmth and shelter and exposed to a pitiless storm. At the bottom of a nearby crevice they could hear the waters of a small stream, even down, past 20 feet of snow, they lowered themselves. The foot-deep water was warmer than the freezing snow above. They began to walk back and forth to keep warm. Here they were somewhat sheltered and, always moving, kept blood circulating until Bain collapsed. A snow bench was carved in the snowbank and the unconscious form placed there until its slender thread of life was severed.

At grey dawning the three survivors pushed on again. Their pilot dead, they knew not the road; all traces of any trail had been obliterated. Late in the day they saw a blazed tree, then another. Renewed energy came with the knowledge they were on the trail again. A final effort carried them over the final stretch to the safety of Peavine.

Major James H. Whitlock led 1852-53 evacuation of Fiddler's Flat on Nelson Creek.

James H. Whitlock

Youth predominated in the early-day mining camps. That was shown during the career of James H. Whitlock, a 23-year-old miner.

In the saga of the winter of 1852-53, Whitlock had led the evacuation out of Fiddler's Flat on Nelson Creek.

Whitlock was born in Illinois. He came overland to the gold fields and arrived in Plumas County early in 1851 after first trying his luck in the southern diggings.

In mid-January, 1853, young Whitlock, who nevertheless was the youngest man in the Fiddler's Flat region, was chosen leader of a company preparing to escape the snows. Of the eleven men who followed Whitlock, three perished. Members of the party became separated. It was Whitlock who organized a rescue party and returned to lead the stragglers out.

Whitlock's heroism was recognized by the voters of Plumas County during the elections of 1854. He was chosen county surveyor on the Whig ticket.

The Snow-shoe Era was then in its infancy. Whitlock, for the next few years, was cutting strange tracks through the mountain snows.

Whitlock's adventurous spirit next carried him into army life during the Civil War. He raised a company of men and was elected Captain. They were mustered in as Company F, Fifth Infantry, California Volunteers. They well may have been the nation's first ski troopers if such had been needed, as they all came from the Lost Sierra.

Whitlock later was ordered to Arizona, where for a time he commanded a post in Tucson. For gallant conduct in a battle with the Apaches in March, 1864, he was brevetted major. Whitlock returned to Plumas County following more than five years of army service.

He resumed his political career, reaching the apex in his election to the California legislature in 1877 on the Republican ticket. From such character-forming experiences did the Sierra's race of snow-shoeing residents evolve.

Interior of a miner's cabin in the early 1850's. Figuring on the value of the gold dust cleaned up on the day. From the California collection of Phil B. Bekeart.

The winter of 1852-53 also saw harrowing experiences on the southern flank of the Lost Sierra. The then burgeoning town of Downieville proved no exception. Snow began to fall somewhat earlier than usual along the North Fork of the Yuba River. No one was prepared for it. Merchants had not received expected supplies. Pack trains had been readied for the trip down below. After three days of steady snowfall, with no signs of letting up, a panic developed and miners by the score began leaving. Next, of course, came the gamblers, then the citizens.

Finally the merchants left. Giving up all hope of getting their trains down the mountains, they hit the trail on foot.

An account of that winter, written by an Argonaut, is to be found in the files of the Downieville Mountain Messenger. It was written a few years later. Here is what it had to say:

"There were seven of us who concluded to take our chances and stay as the merchants gave us the privilege of using anything we needed. Other parties turned over to us whatever eatables they had on hand. Before New Year's we had got away with everything except about a quarter of a barrel of bony corned beef, which was our sole diet and that sparingly dealt out.

"The snow was 10 feet deep. We had drifted tunnel passages through it to several of the stores in vain hope that some crackers or a bit of flour, meal or something might be found to eat. Finally, we bethought ourselves of a patch of rutabagas on Jersey Flat. It was about 40 rods from our quarters and we decided to get there. Where the snow was hard we tunneled, opening places along the route to throw out the snow. About 40 rods of this work brough us to the log across the North Fork. We shinnied over this, and as the patch of turnips we were after was near the river, we soon got at them by tunneling under the snow. But the next thing was to get them back across. This was soon settled by one of us going back to our hotel and bringing back a sack and long ropes. We filled the bag and securely fastened it and, taking it to the crossing, carried the end of the rope across and all hands pulled it over.

From this time on we lived strictly on a root and vegetable diet. Boiled, baked, raw and in chunks or scraped to a pulp, they saved us from starvation.

"How did we pass the time? Well, we read all we could find that had reading on it. Told all the yarns we could think of, sang all the songs, played cards, and of course we had about $800 in dust among us. When one of us was busted we loaned the stake and the next day probably he would have it all. But at the end, each had his own dust.

"By the middle of February the snow had so far subsided that Tom Driver and Dave Sconnars, two old chums, came down from Galloway Hill, bringing each a 21-pound sack of flour for which we gladly paid them $3 a pound."

A miner's jingle recalls that winter:
"Down went McGinty, but he didn't go far;
Still he's snowbound, down at Goodyear's Bar!"

The Lost Sierra had been overwhelmed from the sky, but it was not to be abandoned like ancient Pompeii. Unlike those who 18 centuries earlier had been buried by the wrath of Vulcan, who then dug down to recover their belongings to flee the region forever, the Argonauts of the Lost Sierra, prayed to Ullr, the skiers' god. Their prayers were answered with the arrival of snow-shoes.

References:

Illustrated History of Plumas, Lassen Sierra counties, by Farris & Smith, 1898.
Memoirs & Notes, Albert Dressler
The Mountain Messenger, Downieville.

3

Whipsaws For Spruce

> *"The ancient carnivals of Rome were not capable of producing a greater sensation than our modern and deservedly popular snow-shoeing."* - THE MOUNTAIN MESSENGER, DOWNIEVILLE.

The use of snow-shoes spread rapidly throughout the mountains during the late eighteen fifties. Not only was the use of the long boards a necessity, but from them evolved a new American way of life.

early snow-shoes

Throw-irons and whipsaws were pioneer tools used by the Lost Sierrans to strip the spruce (Douglas fir) billets from which the first snow-shoes were hewn following the introduction of the original Norwegian-styled model skis.

The pioneers arrived with these woodworking implements. Based on research by historians such as Albert Dressler, some foundation can be laid for the speculation that during the winter of 1851-52 the very first crude snow-shoes used in Plumas and Sierra counties were made at Rabbit Creek, a mining town later to be called La Porte.

The new-fangled snow-shoes did not spread into wide usage immediately. Neither did the barrel staves used by Hamilton Ward and James Murray. Those two discoverers of gold in 1850 at the head of Little Grass Valley near La Porte were the first reported to have used over-the-snow equipment. Historians report they fled the snowbound region that winter on their barrel staves.

Early snow-shoes were undoubtedly very crude affairs. Perhaps they were merely flat boards attached to the feet by ropes and only used as extreme necessity required.

Prior to the introduction of true "Norway skates" in the mid-eighteen fifties, as the pioneers' crude wooden snow-shoes were first called, folklore tells of competition from the Canadian or Indian; snowshoes of various web designs. These slow, rawhide-webbed trampers were soon discarded as improvements appeared on the novel, fast

Whipsaws were used used to make planks and beams out of logs. A Plumas County oldtimer mans a whipsaw, above a saw pit, cutting a beam to size as his helper looks on. (Courtesy of the Plumas Historical Museum)

wooden snow-shoes. Up to today the web-styled snow walkers remain more or less of a curiosity throughout the Sierra.

Who whipsawed and throw-ironed the first tree out of which a pair of wooden snow-shoes were made?

Perhaps Henry Harrison Mason, who arrived with his wife and two sons, John and William, in the Rabbit Creek region on Nov. 5, 1851. It is recorded that Masons' belongings included a whipsaw, a crosscut saw, nails and hammers, according to historian Albert Dressler. That is exactly what was needed, not only to build a log cabin, but to fashion snow-shoeing longboards as well.

If Mason first whipsawed a snow-shoe billet the idea might have come from the memorabilia of the famous French explorer, Robert Cavelier, commonly called the Sieur de la Salle, since Mason's wife was a direct descendant of Cavelier.

The Mason family came to the Lost Sierra in a wagon train captained by John La Salle, who had first come to California as a fur trapper. He had explored with Jim Beckwourth, a mulatto guide, when the latter discovered

Throw-irons, used to split logs (Courtesy of the Plumas Historical Museum)

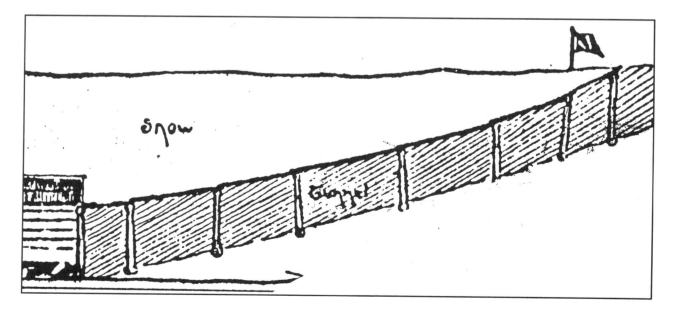

This sketch taken from California Illustrated Magazine, February, 1894 issue, shows how miners got in and out of their homes during the winter of 1989-90 in Plumas County.

the mountain pass which bears his name at the eastern entrance of the Lost Sierra.

The earlier La Salle, first to pioneer in the New World, came to French Canada and explored the Great Lakes region in the seventeenth century. Later he attempted to colonize Louisiana and was murdered by enemies in 1687 at the age of 44 years after having spent some 20 years in exploration and discovery.

That skis, also known by many other names, long have been used in isolated parts of North America has been assumed by many story tellers. That some tribes of North American Indians may have been using wooden snowshoes to glide instead of walk across the snows is not beyond the realm of possibility. Such a statement may bring howls of anguish from some historians.

But the Vikings reach these shores before Columbus. What are believed to be Viking relics have been uncovered in the Great Lakes region. And it is a known fact that some type of ski was displayed in use on the slopes of Mount Royal in Canada's Province of Quebec during the Winter carnival of Courier de Bois during the latter part of the eighteenth century.

Onion Valley and Pilot Peak taken from Hutchings California Magazine, 1860, captioned: "During the winter of 1852 and '53 snow fell in Onion Valley to the depth of twenty-five feet, entirely covering every building in it. The few houses shown in the engraving were all that withstood the immense weight of snow...and there were no less than 13 hotels, besides stores and other buildings there before that winter storm."

Whatever the answers, we like the idea that the Masons whipsawed out the first snow-shoes in the Lost Sierra.

Let us follow them into the birthplace of snowsports.

John La Salle, who was to lead the Mason family and others overland, was among the Hudson Bay Company guides who came to California in 1831 by way of the northern route. He had trapped in what are now known as Lassen, Modoc and Plumas counties, where furs were then plentiful.

Following the discovery of gold, La Salle returned to the States in 1849 with the intention of organizing a company of emigrants. The work of preparing the wagon train required several months. The emigrants, Masons included, left St. Joseph, Missouri, April 5, 1851 and forged westward along the main blazed trail.

The company split into three sections when it reached the Truckee Meadows. That's the spot where Reno, Nevada, stands today. Some headed toward the Sierra crossing at Emigrant Gap, today known as Donner Pass. Others picked the trail toward Henness Pass near the headwaters of the Yuba River. LaSalle, the Masons and others turned farther north and continued westward by way of Beckwourth Pass, trekking through Johnsville in Plumas County.

Here they cut the first wagon trail through to where Gibsonville now stands. Then they made their way to Rabbit Creek, trudging straight through the heart of the Lost Sierra.

Gold fever was high in nearby Spanish Diggings and Mason decided to try his luck. Food and materials, including the whipsaw and other tools, were left behind by La Salle for the Mason family's use.

At Spanish Diggings the Mason family was welcomed by the entire population of nine men, including Col. R. H. Rose, Pop Mendenhall, Jacob Gregg, Ben Colt, John Horn, William Horn, John Irish, James O'Brien and a "Colonel" Cellars.

The following day Mason chose a homesite on Rabbit Creek. Using his tools and aided by several of the encamped miners, he began construction of the first house in the region.

Winter was setting in by the time the job was completed. The other men hastened to house themselves in similar fashion. However the snow came too early and the best they could do was to erect shed roofs over their tents. The snow came quickly in the latter part of November and by Christmas almost all the crude shelters in the Spanish Diggings alongside Rabbit Creek were out of sight under deepening drifts.

The Mason family had the only protection which could be termed a dwelling. It included four rooms, a kitchen, a storage corner, a bedroom and another section used as a catch-all. There was a stone fireplace within and furs, guns and tools were scattered about. In one corner the Masons' dog was nearly always found. It was the only domestic animal in that region. Another corner contained a barrel of sauerkraut. Later on there would also be a barrel of whiskey kept in the corner.

The liquid spirits moved in along with John Irish, when that hardy soul's nearby tent collapsed under the snow. After Christmas, the celebration of which was confined to sauerkraut eating and alcohol toasting, the windows had to be boarded up. There was 15 feet of snow on the level and the canvas window lights were on the verge of bursting inward. The Masons' candle supply was soon exhausted. The only source of illumination was the fireplace.

Shortly after New Year's food began to run out. Mrs. Mason carefully doled out her sauerkraut. There remained some flour, beans and roots, all of which eventually were turned into thin soup.

March 17, the occasion called for Irish to get himself considerably liquored up. He suggested the dog would make at least one good meal. To save the dog Mason bethought himself of his whipsaw and soon was outside with some kind of "snow-shoes" on his feet and gun in hand. When he returned several hours later he was in possession of a rabbit for the stew pot. That is the legend of how snow-shoes came to Rabbit Creek.

Early in April 1852 men began to flock up from the valley diggings. The miners, in order to protect themselves from trespassers, staged a meeting, April 25, to fix limits on mining claims. They elected an alcalde, naming Colonel Rose to that office.

The first substantial structure actually built at Rabbit Creek was the Rabbit Creek House. It opened its doors for business in the fall of 1852. It included a trading post and saloon.

Rabbit Creek was soon destined to become the principal commercial and social center for all the nearby

Young America's Thoroughbred snow-shoers Start the Race, La Porte, Circa 1910. Crowds and excitement followed snow-shoe races as shown in this scene with miners poised at the starting gate.

diggings, including Spanish.

In 1853 other houses and stores were built and the Masons found themselves in the midst of busy frontier community life.

Rabbit Creek grew rapidly and by the winter of 1857-58, 3,000 men could be assembled in an hour's time if necessary. Gambling, drinking and eating engaged the establishments full blast night and day.

During the previous summer the citizens had changed the community's name to La Porte. It was now approaching the zenith of its prosperity. It boasted three hotels, half a dozen stores, a ten-pin bowling alley, 14 saloons, two dance halls, two churches, a snow-shoe company. The winter of 1857-58 featured more damned snow than anyone wanted.

winter sport begins

The great snow depth was neither good for business nor the morals of the miners. Snow-shoes now had come into general use and those who catered to the recreational needs of the idle men knew that something extraordinary was needed to while away the tedious months during which all mining activity annually was forced to a virtual standstill.

So snow-shoe racing was instituted during the winters of either 1855 or 1856 in Onion Valley and Poorman's Diggings according to old-timers' tales. Unfortunately, there are few contemporaneous accounts of snow-shoeing during the fifties. The mountain newspapers of that period were conveniently small in size and of necessity almost immediately consumed as fire-starters or found other domestic economy uses.

There are some reports of ski racing taking place in

Mail carrier with his pouch, carried mail from the valley post offices to mountain residents in the "Lost Sierra." Photo from The Californian Illustrated Magazine, February 1894.

saloons for the downhill race track, but they took their bars with them. It was a new sport for young men as seen by early day reporters for the Downieville Mountain Messenger. They had a lot to say about the races.

A Mountain Messenger reporter rhapsodized about the early downhill skiing:

"The ancient carnivals of Rome were not capable of producing a greater sensation than our modern and deservedly popular snow-shoeing. There was the greatest fairness and good feeling manifested on both sides. Nothing occurred to interrupt or mar the proceedings; no excessive indulgence, no liquor, or no perceptible effect of indulgence; no pugilistic displays. I contrasted it in my mind with horse racing in the East, where I have known of several arrests made for the last offense, and which in this country, on the borders of civilization one might say, and where the moral influences are so few, I know not what cause to assign, unless, indeed, the early impressions received in childhood or early manhood still remained, and who in youth were impressed with the impropriety and wickedness of indecent conduct.

"I was saying, when I slid from my subject, that everyone seemed pleased and the feeling was contagious, inasmuch as I looked upon the whole scene with the profoundest admiration, the contestants included, but only as such from an artistic point. Perhaps the free mountain air had a tendency to inspire a feeling for the beautiful in nature or art. The snow-shoers lent a lustre to the scene and made us more susceptible and more appreciative, and the ball each evening was largely attended by Jew and gentile. Politics nor religion caused no line of demarcation in their social enjoyment: the pleasure seekers were bound to fill the hall without regard to race or color."—The Mountain Messenger.

The use of snow-shoes spread rapidly throughout the mountains during the late 1850's. Not only was the use of the long boards a necessity, but from them evolved a new American way of life. Among the early newspaper accounts of community snow-shoeing activities was the following from the March 26, 1859 Quincy Plumas Argus:

"Having seen nothing of the Argus for some time, I think it best to write to you that Poorman's has not gone in yet, although a stranger would think it was about to go under this winter from the quantity of snow which has been falling here.

"This has been the hardest winter within the knowledge of the oldest inhabitant. It is estimated that about 25 or 30 feet of snow has fallen at different times this winter. The snow now lies from 8 to 10 feet deep, but it is not thought much of, for at Onion Valley, two miles from here it is 12 or 15 feet deep.

This sketch was printed on the outside of envelopes used for mailing by Zacharia Granville's Express Company, La Porte, in 1857.

"It may be a matter of wonder to some of your readers how people get about where there is so much snow. It is the easiest thing in the mountains. Nearly all have Norwegian snow-shoes. They are about nine feet long, four and one-half inches wide, shaved thin and turned up in front like a sled runner. By fastening them to the feet about the middle of the shoe and with a pole in the hands for a balance, a person can run over the light and new fallen snow at railroad speed." — The Plumas Argus.

Emissaries of the new sport came mostly in the form of snow-shoe mail and express carriers. Soon each community had its own snow-shoe manufacturers. All residents, miners, merchants, hotel and saloon keepers, gamblers, all other professional men, ladies, newspaper editors and ministers of the Gospel were enjoying winter recreation on the snow-covered slopes.

delivering mail

During the very first three or four winters in the Lost Sierra the pioneer mail and express carriers had great difficulties. At that time snow-shoes, the snow skate variety, were either unknown or only used by scattered Scandinavian settlers. The luckless messenger had to flounder through the snow as best he could. Then the Indian or Canadian snowshoes were introduced. With these on his feet and his bundle of letters on his back, the expressman was able to make fairly good time over the snow when it became too deep for animals. However, this proved too slow for several of the energetic carriers and various types of longboard snow-shoes were introduced during the winters of 1853-54 and 1854-55.

The following is quoted from The Californian Illustrated Magazine, February, 1894:

"To the inhabitants of the snow buried towns of Plumas County, there is no visitor hailed with greater delight than the mail carrier. During the severest storms of winter, communication is maintained for long periods only by snow-shoers, who carry the mail upon their backs. They are always expert snow-shoers and attempt but seldom to handle more than the letter mail, leaving the papers and bundles to be brought in upon sleighs later on. It often happens that no man, no matter how strong he may be, can face the storm and carry to the people of the next town their mail. As soon as the weather clears up, parties of volunteers start out and "trail" is broken through. At each turn, where a tree may be conveniently situated, they nail a piece of board or shingle in order to indicate the exact location on the road. In summer when all the snow has disappeared, these guide-boards act as reminders of the great depth of the winter's snow."
— H. G. Squier

dog sleds speed mail

Along with the snow-shoes another unique innovation surfaced to speed the express and mails. It was a development that sprang from the fertile mind of *Fenton B. Whiting, a youthful native of Virginia, who had tramped through the snows of February in 1851 to reach Onion Valley.

During the winter of 1854-55 young Whiting became an express snow-shoe carrier on a route between Junction Bar and Bidwell's Bar. Later he organized the Feather River Express to carry mails, passengers and supplies between Buckeye and Meadow Valley. Like all American boys of good education, Whiting had read stories of Polar exploration. It now occurred to him that he could adapt sledding and dogs for use on his express system. He procured three large, strong and intelligent dogs of the Newfoundland and St. Bernard breeds. He broke them to work in harness. He had a special sled constructed, the wooden runners of which were wide and grooved on the bottoms, like snow-shoes, to prevent them from slipping sideways. The trial trip was a magnificent success. Meanwhile on other routes, among the region's maze of blazed trails, the snow-shoe mail carriers continued to pole along.

Whiting & Co., "Feather River Express", was organized in 1856, by Fenton B. Whiting, prominent Gold Rush Era expressman. He pioneered the use of dogsleds to deliver mail to miners. Plumas County Museum photo.

even horses wore snow-shoes

Dog sled express went out of style, however, when yet another innovation in winter travel was introduced in the early sixties. That was the season they put snow-shoes on horses. Visual evidence of the practice can be found to this day in almost any abandoned livery stable in the Lost Sierra.

At first horses' snowshoes were made of wood, but these horse snowshoes were quickly replaced with metal shoes because snow had a habit of sticking to wood no matter how expertly the drivers doped or waxed the bottoms.

Thinner plates of steel then were adopted. They used rubber lining on the bottom, for which the snow had no affinity whatsoever. The horse snowshoes were 9 to 11 inches in diameter, and slightly streamlined or tapered to the rear. They were fitted to the horses' hooves by setting the caulks of the horseshoes through holes in the plate of

33

the snowhoes. They were fastened firmly with screws and straps.

The snowshoes had to be fitted individually to each horse, as their own shoes varied in size. It took a good man two hours to place the snow-runners on a four-horse team. When first attached, some horses cut themselves about the hocks, but most learned to spread their feet to avoid the problem. Some animals became experts at once, while others seemed incapable of learning the art no matter how expert the instructor. Horses that became accustomed to the snow seemed to use intelligence and judgment. They would battle the fleecy white carpet draped over the mountains as could be expected of the most artful human snow-shoer. To this, the best of the Lost Sierra's many expert story tellers, will testify to their dying days.

snow-shoes and matrimony

Meanwhile in La Porte John Mason had reached the ripe old age of 12 years and hired himself out as a snow-shoe carrier; for the town's leading bankers, John Conly and Aleck Crew. His career began over the short routes to Spanish Digging, Port Wine and St. Louis. He continued the job until his marriage in 1873.

As he progressed from boyhood to young manhood, Johnnie Mason's snow-shoe trips extended to such far-flung areas as Howland Flat, Gibsonville and Onion Valley. During his messenger career he carried hundreds of thousands of dollars in gold dust, coins and currency over these routes.

Young Mason was held up but once during his perilous work. The incident took place in 1860 when a white man and three Mexicans jumped him in a brush-shrouded ravine. He was on a return trip, and the bandits' thorough search uncovered only two dollars. At another time while carrying $11,000 to Gibsonville, he suddenly came upon the freshly murdered victim of highwaymen, throat cut and pack contents scattered about.

Such experiences were routine for mountain messengers during the sixties. However, one snow-shoe trip was not routine. Mason snow-shoed to Morristown on a special mission for Banker Conly. He was forced to put up overnight in the local hostelry. In the morning an old friend, Webster Walker, invited Mason to the Walker home for breakfast. There he met 15-year-old Laura Walker, who was frying pancakes. Needless to say they were of the finest kind. When Mason returned to La Porte he persuaded the bankers to send him right back to Morristown. Many other trips were arranged and after a four-month courtship Laura reached her sixteenth birthday and consent was given by the Walkers. The next day Mason hiked down the mountains to Downieville; and procured a license. On July 3, 1873, John Mason and Laura Walker were married by the Rev. William Gordon of Downieville in the Methodist Episcopal Church at Morristown.

In October of 1880 the John Masons moved from Rabbit Creek to Downieville, where their two hearts beat as one for more than 50 years. He always asserted that he had intended to leave the snow county in the spring of 1881 and explained: "When I came down I could not pay the toll across the bridge below town which was ten cents a head. After I stayed a year I owed so much they would not let me go."

The veteran snow-shoer was elected constable in 1884 and named deputy sheriff under a Republican administration. Then followed ten years of law enforcement under the Democrats and another eight under the Grand Old Party. He purchased a house from Major William Downey, for whom the city of Downieville was named. He was a snow-shoer and loved it.

In 1930 he sent the following greetings to an old Rabbit Creek acquaintance:

"I am now, and have been, for twelve years, Justice of the Peace. Any romantic couples who wish to be married, come forth!"

Snow-shoes and hot cakes did it.

References:

California's Pioneer Mountaineer of Rabbit Creek, by Albert Dressler, 1930.

William B. Berry, Interviews with Oldtimers; Lenny O'Rourke, Albert Dressler, Col. J.F. Mullen, A. Pock, Harvey and Dorothy Egbert.

The Quincy Plumas Argus, 1859

The Mountain Messenger, Downieville.

The Californian Illustrated Magazine, February, 1894.

Cleve (Cleveland Henricks O'Rourke), mail sleigh driver, gives us a closeup picture of what snow-shoes on horses looked like. This picture was taken at La Porte, Circa 1909.

4

Quicksilver Charley

"We have learned that a club of 25 snow-shoers in St. Louis recently challenged an equal number from Pine Grove to run a race for a champagne dinner." — *(First report on snow-shoe racing,)* - THE DOWNIEVILLE MOUNTAIN MESSENGER, MARCH 14, 1863.

Charles W. Hendel with a group of local lovelies whose company he always found enjoyable. He earned the nickname "Woodbox Hendel" when a strapping lady snowshoer responded to his advances by tossing him in a nearby woodbox. W.B.B. Collection.

Each regional snow-shoe manufacturer designed an individual tip. These were the makers' trademarks. Many were of artistic merit. Through them, all spectators knew which pair of longboards came from La Porte, Howland Flat, St. Louis, Poker Flat, Gibsonville, Port Wine, Onion Valley or Poorman's Creek, as the case might be.

wood box hendel

Charles W. Hendel of Old St. Louis and La Porte had two pseudonyms. As "Quicksilver Charley" he was the most eloquent chronicler of the Snowshoe Era in magazines and newspapers. Until his death, he was also known to his intimate friends as "Wood Box" Hendel.

Drawing from The California Mining & Scientific Journal, 1874, showing racers coming down the course, as envisioned by the artist, who did not show them racing abreast as was customary by that time. Miners in the foreground are shown enjoying their favorite beverage. The crew at right is "doping" longboards in preparation for the next race.

Through the writings of Hendel, who came to the Lost Sierra in 1853, sufficient evidence can be found for the belief that true Norway-style "snow skates" were first introduced in December of that year. In an article written for the California Mining and Scientific Press, published during the winter of 1874, Hendel had the following to say:

"So great have been the improvements made in racing shoes, during the past few years, from the original style, first introduced (approximately) 20 years ago, that they now appear to have reached perfection."

That documents 1854 as the date longboard snowshoes appeared.

Hendel was well qualified to approach the subject of snow-shoeing from a scientific viewpoint, having graduated with honors from the Zchocko Technic Institute, Dresden, Germany, in 1850. He was to apply his mind and energies to improving snow-shoes during the many winters he spent in Plumas and Sierra counties until his passing at La Porte in 1920.

No phase of snow-shoeing escaped Hendel's keen eyes and mind. He carefully noted the measurements and design of the longboards which had been developed to carry man at speeds of up to mile-and-a-half-a-minute.

They were from 10-1/2 to 13-1/2 feet in length, varying in weight between 13 and 17 pounds per pair. The length and weight of the snow-shoes was predicated on the weight and height of their rider. Widths varied from 2-3/4 to 4-1/2 inches. In all instances they were slightly wider on the front part than on the rear.

Racing snow-shoes were fluted with a 1-1/2-inch-wide and 3/8-inch deep groove along the bottom. On top of the shoes, a little back of center, there was about 18 inches of wood left flat. Toward the front they were shaved and planed until they tapered sufficiently to make the points springy. There was considerable wood left behind the center to the rear end, which made for proper balance. Little or no spring was required on the back part. The object of that was, that in running over rough places, there would be no sudden jerk, endangering the equilibrium of the rider, as his speed hit 60, 70, 80 or even 90 miles per hour. They had a tendency to "buck" when going over uneven snow. Riders often found they were as uncertain as all other things are here below.

The snow-shoes from different regions varied in one particular part. Each regional snow-shoe manufacturer designed an individual tip. These were the makers' trademarks. Many were of artistic merit. Through them, all spectators knew which pair of longboards came from La Porte, Howland Flat, St. Louis, Poker Flat, Gibsonville, Port Wine, Onion Valley or Poorman's Creek, as the case might be.

Details of longboard bindings are shown by the author, Bill Berry, left, and Wendell T. Robie at the entrance of the Auburn Ski Club's William B. Berry Western American SkiSport Museum at Boreal Ridge in 1969. photo by Chapman Wentworth

The daring rider of racing snow-shoes stood a little back of center of the snow-shoe. His feet were secured by toe straps of strong leather or India rubber belting

This 1890 photo, taken at La Porte, shows how racers lined up for a "Free-for-all-style," start. From left, J.J. Bustillos, Ed Mullen, Neal Mullen, Mike Bustillos, Dopemaker Tom Larimore and Coach Jesus Bustillos. W.B.B. Collection.

which was fastened to either side of the shoes and laced where they met the foot. The toe of the snow-shoers' boot was put into the straps back to the ball of the foot, in the hollow of which a small block was placed crosswise to prevent it from slipping out of the straps if the heel was raised.

That the racing bindings provided maneuverability was pointed out by Hendel. Comparing the "snow skates" to those used on ice, he said, "To some extent the same evolutions are practicable, such as allowing the points and curves to describe a circle." He was quick to point out, however, that, of course, "they cannot be turned so easily or quickly as skates, but they still are easily managed by experts."

However, The sine qua non of Snow-shoe Era racing was "dope." Today we call it ski wax. Dopemakers in the Snow-shoe Era developed a science that may easily have produced a product superior to today's commercial wax formulas.

Hendel described it as "the material used to lubricate the bottom of the shoes and cause them to glide over the snow, as an axle is lubricated, to cause the wheel to revolve easily. The object is to counteract friction as much as practicable."

The temperature of the snow, he pointed out, is as variable as that of the atmosphere, and "for every temperature of snow a different kind of dope is required". Every snow-shoe racer, or dopemaker as the clubs' coaches were called, had at least half a dozen recipes for compounding dope. This substance was sometimes called, "Sierra lightning," or "Greased lightning."

Hendel wrote: "There is one dope for cold snow and one for warm or damp snow, as it is called by experts. There is another for dry snow and one for wet; one for hard and one for soft; one for forenoon and one for afternoon; one for extreme cold or frozen snow. For new, dry snow, there is still another kind required. Some go as far as to have a different kind for every hour of the day. For moist snow the dope is soft. It is made harder for an increase in temperature, up to the frozen, when a hard dope is required."

Great skill and ingenuity were required in manufacturing dope, much depending upon the time element necessary for amalgamating the various ingredients. Some recipes needed only a light simmer to blend, while others required a good deal of boiling.

Hendel emphasized that gum, beeswax, rosin, sperm candle and other such common materials (Modern skiers take note!) made only an "inferior quality of dope, used only for travelling purposes."

The bottoms of all snow-shoes were highly polished. Pine tar was burned and rubbed in until a full, mahogany-like finish was obtained. This process hardened the wood and attracted heat when exposed to the sun, a desired effect in applying dope for both traveling and racing.

Lightning dope for racing was described by Hendel:

"It is manufactured from spermaceti, Burgundy pitch, Canada pitch, balsam of fir, spruce, cedar, Venice turpentine, oil of cedar, pine, hemlock, fir, spruce and tar, glycerine, Barbary tallow, camphor castor oil and many costly drugs known only to those who make it a specialty, and its manufacture a secret. Varnish or any other polished material is useless. Nothing but the scientific preparation will do; for it is a common saying among snow-shoers that "Dope is King." The dope, in order to be good, must possess two qualities: First, it must be sticky, so that it will adhere to the shoe; second slippery, so that it will glide over the snow. And, strange as it may seem, they have attained such a degree of perfection in making this compound, that a snow-shoe prepared with it and placed by the side of one with the bottom finished with polished steel, would so far outrun it as to make it no race at all."

Charley Hendel was one of the great sportsmen of the region and none excelled him as a recreational snow-shoer. He witnessed the early-day snow-shoe races in the mid-fifties. He became a member of the Howland Flat Snow-shoe Club when that organization was formed in 1866 and later joined the Alturas Snow-shoe Club at La Porte where he was present for the finals of squad racing when the last tournament was held in 1911. Only Hendel and a handful of others spanned the entire period.

Around 1859 the contests changed to "free-for-all style," with two or more miners starting together. As the number of participants increased, the skiers were placed in groups, each group representing a mining camp.

On Feb. 7, 1861, the Sacramento Daily Union reported:

"Great big men, extremely small children and delicate looking females ascend La Porte's Sugar Loaf Mountain, and how they came down. Everyone who has the time to go out in the evening for an hour or so, and every person who can buy or borrow a pair of skis is out on the mountain, and they appear to enjoy the sport."

The same paper reported on March 4, 1865, that Lincoln's birthday was celebrated by a two-day ski meet on Mount Pleasant Gibsonville. Winner of the principal race received $50.

The March, 14, 1863, issue of the Downieville Mountain Messenger reported under the headline:

"SNOW-SHOE CHALLENGE"
"We have learned that a club of 25 snow-shoers in St. Louis recently charged an equal number from Pine Grove to run a race for a champagne dinner. The challenge was promptly accepted and the course laid out and the race will take place next Tuesday. Each club is to choose 11 of its best runners to compete in the race. It is our impression the boys will have a pleasant and jovial time.

On March 21, 1863, the Downieville Mountain Messenger reported under the headline:

"THE SNOW-SHOE MATCH"
"The snow-shoe race between the St. Louis and Pine Grove clubs, which was to have come off this week, has been indefinitely postponed in consequence to disagreement between the parties regarding the terms and particulars. There is four feet of new snow, accompanied by high winds which howled dismally through the dark pine forests which hide our hills."

The inter-club tournament was finally run off on All Fools' Day, with a $150 purse at stake. An excerpt from the Mountain Messenger's lengthy account stated:

"The ground to be run upon and the place was agreed upon by the judges. The hill is very steep and upon it no ordinary rider can stand upon his shoes.

"The distance was 2,000 feet, but I must hasten to give an impartial and unprejudiced account of the races as they were run. The first race was run by Tom Kelly of Pine Grove and John Berman of St. Louis. Both came down the dizzy grade in pretty style, neither falling and at a rapid rate. Kelly passed the stakes ahead perhaps by 50 feet. The second race was run by Edward Danfer of St. Louis and J. Barrett of Pine Grove. This was the most closely contested race of the day. Both rode with breathless speed and confidence glowing upon the faces of its contestants, but Danfer with the speed of a winged messenger passed the poles in most envious style, slightly ahead and winning."

Kelly and Danfer may have been the first racers ever to win in an organized club tournament. While there are numerous accounts of championship races in previous years, the results of the tourney between the St. Louis and Pine Grove boys provided the first specific documentation of an actual snow-shoe club race.

While the Howland Flat and Alturas ski clubs can be credited with having been the world's first clubs formed for the specific purpose of placing snow-shoe racing on an organized basis, they were antedated by many other snow-shoe clubs, the members of which were primarily interested in the social angle, but engaged in racing as well. We don't know which of the many mining camps of Plumas and Sierra deserve the honor of starting the world's first ski club. However, Charley Hendel, as the St. Louis correspondent of The Mountain Messenger, was keeping the editors well informed of lively ski club events prior to 1866.

Organization of the Howland Flat and Alturas clubs three years later took the sport out of the hands of committees or social organizations and placed it upon a formal association basis. Uniform rules were set up and other details worked out by the new clubs.

Racing tracks were selected and cleared of trees, shrubs and other obstructions. The courses were between 1500 and 2500 feet in length, with an angle of descent of 15 to 35 degrees, always in as straight a line as possible. The winning poles were set on comparatively level ground to give the racers a chance to turn out or, for the middle men, to "brake up," which was accomplished by dragging their poles behind and bearing heavily astraddle them in a sitting posture.

There was an entry fee of $1 for each race. The competitors were divided into "squads" of four, six or eight, generally representing different towns. The winners of each squad race made up other squads in like manner and in round-robin fashion until the finalists squared off to decide the actual purse winner.

The following colorful description was penned by Hendel:

"Time being called, the racers take their places, side by side, all their boot fronts in line. They hold back with their poles, on the bottom of which a flat round button of wood is fastened.
"The instructions are given:
"The first man through the poles, on his shoes, or one shoe, wins. Anyone starting before the tap of the drum cannot win.
"The drum is tapped out of their sight: at the same time a red flag is dropped to inform the time keepers and judges on the lower end that the racers have started (time kept with stop watches), and they shove off with their poles, poling for dear life and giving as many strokes as they can to accelerate their speed. There is a breathless silence, from the

start, for a few seconds. Sometimes one of several fall on the start by striking their shoes in poling. Now they are running 40 miles an hour, now 60 in a second or two they are going at the rate of 70 or 80; here one runs over a rough place and loses his balance, his shoes fly in the air and touch another. Away he goes and throws another, and heads, legs, arms, poles and perhaps broken shoes are turning in 20 and 50 foot somersaults, amid a cloud of snow, raised by the current of air produced by their fall and velocity. If the shoes are not broken, they shoot off in the air like an arrow, or like a riderless warhorse in battle, over the snow field, and leap often several feet in the air, until they run against a tree or into a ravine, where the get stuck on the other side, if they do not go again over the next hill in the snow bank."

Free-and-easy dances followed each day's racing. The major tourneys lasted three to five days. The snow-shoers' conversation turned on nothing but the state of the snow, dopes, pretty riding, gymnastic performances and times made on other occasions.

"In no place but in California," Hendel once told such a gathering, "can so many men meet, contesting for prizes and the reputation of so many towns, and part in the utmost friendship."

Charley Hendel was a bachelor who liked to dress well. He was sociable and particularly fond of children. Every summer he would go away for a brief vacation. One of his last was to the World's Fair in 1915. It lured him down to San Francisco for several weeks. He was about five feet, five inches tall, with a wiry, athletic figure. He was always on the go, always bubbling over with good nature and good health. It was the propensity to be always on the go and his agility on snow-shoes that earned him the nickname "Quicksilver," a pseudonym he adopted when he penned for the mountain press.

The "Wood Box" handle given Hendel was nailed onto him in another way.

During the early winter at a snow-shoers' dance in Howland Flat, Hendel, along with several others, was paying court to the winner of the girls' race that day. She was very good looking, and needless to say, remarkably strong. He said something she didn't like, whereupon she picked him up and slammed him into a large wood box. Faster than anyone could say "Quick, Silver," he had become "Wood Box" Hendel...and that was that!

Hendel's ardor was thus cooled and his life from 1861 until his death was strictly a love affair with the Lost Sierra. In the fall of 1861 he returned to Germany for a visit with his parents in Saxony, where he had been born, July 21, 1831. In February, 1862, he maintained a correspondence with The Mountain Messenger, which published an extract of one of his letters. It was almost enough to melt the snow in the deepest Sierra canyon:

"Sleigh riding, ball going, singing parties and hunting are now the order of the day. Everywhere they call me the Little Yankee, and you bet I try to be a real one in every manner, shape and form. When I sit in a chair, I must have an extra one to put my feet on. And, by the way, in regard to kissng the pretty girls, I get a little taste of it now, after so long a fasting in California. It appears they like the California moustache very much, although they say Californians must be nasty on account of their great beards. I expect to leave Germany in the beginning of May, so you may look for me some time in June. I assure you I am really home, and lovesick after beautiful California."

Following Hendel's return to Sierra County in 1870, the voters honored him by naming him county surveyor. His transit and level ran the lines for many of the spectacularly scenic dirt roads throughout the Lost Sierra. Several are still in use today.

In 1871 Hendel was appointed deputy United States mineral surveyor, a post he held until his death. It was also in 1871 that he moved to La Porte and in 1879 was elected Plumas County Surveyor.

Shortly before Hendel's passing in 1920, the question arose as to who would receive his historically valuable field notes, maps and other memorabilia of early days in the snow.

The California State Library, Sacramento, wanted the complete collection, which Hendel wished them to have.

However, Quicksilver Charley was a trifle sensitive when any discussion involved his advancing years and he kept putting off drawing up his will until it was too late.

There were no known relatives and the disposition of his affairs was placed in the hands of the public administrator at Quincy.

It was thought at the time there would be no objection to having most of Hendel's valuable collection sent down to the state library, Sacramento. An official representative, Harry C. Peterson, was assigned to the task. He battled a snowstorm to reach La Porte.

"You ain't goin' to take Grandpa Quicksilver's things are you mister?" anxiously queried a seven-year-old youngster of Peterson's.

The youngster's eyes were ablaze with anger that an unfeeling state should so separate the memory of his beloved Grandpa Quicksilver by taking his things away.

Very patiently Peterson explained to the boy that the California State Library wanted Hendel's papers and writings deposited in the great archives of the State of California, where the whole great state could benefit by them.

Peterson then entered the house to find everything just as Quicksilver had left it the morning he died. He had never been ill, so that when he complained to Will Pike over at the store one night that he did not feel well, Pike told him to cover up when he went to bed and in the morning he would find the sun shining. But in the morning when Elmer Primeau arrived with a special delivery letter he found Charley Hendel had reached the final amalgam of time and Quicksilver. He had reached 89 years, one month and 26 days.

Unwashed dishes were on the little kitchen table, the lamp had not been filled. On the great drafting table was an unfinished map, and on a little desk lay a partly written letter. Weather reports and observations were stacked on all sides. Hundreds of books lined the shelves. Scores of leatherbound notebooks, containing field notes of all that mountain country jammed several shelves. The drawers bulged with ore specimens, including many boxes of indexed clippings. In the woodshed stood several pairs of snow-shoes, some dating from earliest times, side by side with many single poles. Up until he was past 80 years of age, Quicksilver had made monthly snow-shoe trips over the 34-mile route to Quincy to attend meetings of the Plumas County Board of Supervisors, on which he served from his election in 1910 to the time of his death.

It was Peterson's duty to go through all these things and pick out what he wanted. He filled one trunk with newspaper clippings. Another he filled with deeds and mining claim notices. Boxes were packed with pictures, for Quicksliver had been an amateur photographer. Also prepared for shipment were the historically valuable snow-shoes and other memorabilia pertaining to the

Skiing was not always serious as seen in this posed photo, taken at Howland Flat race course around 1890, showing snow-shoers tumbled about finish line between the poles and lady snow-shoes in hoop skirts at right. W.B.B. Collection.

Snow-shoe Era.

However, of all the material packed for shipment, little ever arrived in Sacramento. The great trunks and snowshoes were burned when the house caught fire not long afterward. The maps, it later developed, were taken over to Quincy and sold by the public administrator to the highest bidder. Little by little the collection, or that part not destroyed by fire, became scattered. In the state library today there is an unindexed mass of Hendel's papers. In the Plumas County Library at Quincy is what remains of his scrapbook on the Snow-shoe Era. Much of the scrapbook was looted before it found its present resting place.

Had not a storm inconvenienced Peterson he might have been able to take the mass of Hendel's material with him. But the snow falls deep in the Lost Sierra, and the result was that the public has forever lost some priceless Americana.

References:

California Mining & Scientific Journal, 1874, article on skiing, written by Charles Hendel.

"The Challengers," La Porte Union 3/1/1869. Vol 1, Issue No.1.

Sacramento Union, Feb. 7, 1861.

Downieville Mountain Messenger, March 14, 1863 and March 21, 1863.

Flowers still appear each year on Charles W. Hendel's grave at the La Porte cemetery. Other La Porte leaders, like Truman Gould, former general store owner, are buried near by. CW photo.

SNOW SHOE RACES!

FOUR DAYS SNOW-SHOE RACING AT
HOWLAND FLAT,
UNDER THE AUSPICES OF THE
TABLE ROCK SNOW-SHOE CLUB,
Commencing on Monday, March 15th, '69.

PROGRAMME.

First Day. 1st Race. Club Purse of $125 free for all. 2d Race. Entrance money of the day free for all but winner of the first race.
Second Day. Club Purse of $75, free for all. 2d Race. Entrance money, free for all but winner of the first race of this day.
Third Day. Club Purse $75, free for all. 2d Race. Entrance money, free for all but the winner of the first of this day.
Fourth Day. 1st race. Club Purse of $125 free for all. 2d race. Entrance money free for all but winner of first race this day.

Purses for Boys will be made up during the races. Racing to commence at precisely 1 o'clock. All entries must be made before 11 o'clock A. M. with the Secretary. Entrance fee $1. If the weather should prove unfavorable on Monday, March 15th, the Races will be postponed from day to day until favorable.

FREE DANCES DURING THE WEEK
and a GRAND BALL on St. Patricks Night

the 17th inst. at the SIERRA NEVADA HOTEL. During the week of the races the Sanhedrim of the Ancient and Honorable Order of E. C. V. located at Howland Flat, will, by permission granted by the G. R. N. H., have a celebration, procession etc.
By order of the Club.

Sam. Wheeler, Secretary.   **D. H. OSBORN, President.**

See also a scholarly treatise on this subject entitled
"Poineer Skiing in California" by Clamper Bob Power, of Nut Tree.

Table Rock Snow-Shoe club hosts four days of racing at Howland Flat, as advertised March 15, 1869 in The Downieville Mountain Messenger.

5

Stake Racing Days

Both started handsomely, making some half-dozen strokes with their snow-shoe poles in very quick time, that sent them forward with great velocity. Young Lee seemed to be leading his man in gallant style until they came to the most precipitous part of the track, where Metcalf's shoes darted by and led the way to the poles, winning the race by about 15 feet. Time: half a mile in 25 seconds." DOWNIEVILLE MOUNTAIN MESSENGER, FEB. 21, 1863.

Onion Valley, in a 1910 winter scene with Pilot Peak, a popular race site, in the background. The two-story building, at center, left, is the notorious Husum Hotel, a "sporting" house, built in 1860. The photo, taken by Ira Mullen, was contributed by Elizabeth Merian, whose grandfather's barn and chicken house can be seen in the foreground.

...it became clear that the speed of the shoes could be greatly accelerated by giving the proper character to this auxiliary power mixture. We call it ski wax. Back then they called it "dope."

First among Lost Sierra mountain communities to welcome the introduction of snow-shoes were Saw Pit Flat and Poorman's Creek on the east side of the ridge above Onion Valley and Pilot Peak.

Nothing much now remains in that isolated region to show it was once bursting with life. In 1852 Sierra County had a population of 4,855, according to the

The winter of 1879 at La Porte. Women dressed in their Sunday best donned their snow-shoes to go to church and other social functions. Using the second story to enter homes in winter was commonplace. W.B.B. Collection.

California Secretary of State. Downieville had a population of 810.

Mountain peaks towered over the valleys. The special 1852 Sierra County Census showed: "Saddle Peak, height 7200 feet; Table Mountain, height 8,000 feet; Buttes at head of South Fork, 9,000 feet surrounded with quartz leads, limestone excellent, no mineral springs."

snow-shoe competition speeds up

It was in these two, once-famous, roaring mining camps that the claim for the most important "firsts" in snow-shoeing history can be boasted. Credits include the manufacture of traveling and racing snow-shoes, development of dopes, inauguration of actual competition and victorious show-shoe racing championships.

Onion Valley took its name from a wild onion that flourished there in the early days. It provided a relish for the miners. During 1851 Onion Valley became a general trading center for the surrounding area. Soon it featured several hotels, saloons, stores, homes and, of course, a "sporting" house, called the Husum Hotel.

Saw Pit Flat had claimed its name because lumber was whipsawed there. Nearby stood Poorman's Creek, whose first diggings had not made anybody rich.

The earliest snow-shoe manufacturer in this region, and possibly the entire Lost Sierra, was John Porter. He was a Saw Pit carpenter known familiarly as "Old Buckskin." From Porter's wood-turning shop came what were called the "Norwegian improved upon" snow-shoes.

Original models for traveling introduced in that region during the winter of 1853-54 were flat bottomed. They were promptly copied by Porter who later "invented" a groove for the bottoms. This popular feature could check side slip, a real bonus miners enjoyed when they took up snow-shoe racing. The sport took off after miners found longboards could provide fun as well as convenience. Originally Porter chiseled out the grooves by hand. Later, he designed a tool to reduce the labor required for that purpose.

When one miner challenged another to a snow-shoe duel, the attention of the adepts became directed to the production of a concoction that would excel that of all other competitors. Thus it became necessary for the racers to devise recipes and manufacture their own dopes, or employ some expert, under strict injunction of secrecy, to make it for him.

Saw Pit Flat enjoys the reputation of being the first mining camp to succumb to snow-shoe fever. Its first accredited combination dopemaker-racer was a miner named William Clinch. Other competitors included William Metcalf, who was called "Uncle Bill", Peter Riendeau, known as "French Pete" Robert Oliver, nicknamed "Cornish Bob", Napoleon Norman, dubbed the "Little Corporal," and John G. Pollard, called "Grey."

These men's names, the first interwoven into snow-shoe legends, are to be found in old newspaper accounts starting in 1861. Metcalf was a "champion" during stake-racing days in 1861 and 1863. Oliver won the first championship belt in 1867 when the sport was developed into an organized effort. Riendeau was the 1868 winner. Norman's racing career ended in a tragic accident, while Pollard for many years performed as the fleetest.

Original racing rules were comparable to modern downhill ski racing, although the competition was usually limited to two men vying for a posted cash prize stake. A course was posted and the racers rode down singly against time. This type of racing remained in vogue from 1855 until 1858, when the running was turned into "free-for-alls". Usually squads of four riders were sent away simultaneously.

By the winter of 1861, during which Metcalf won the Onion Valley championship, the combination of dopemaker and racer was going out of fashion. The riders at that time were assigning the problems of fast "dopes" to those who made its manufacture a business. Dopemakers became the coaches of snow-shoe racing.

Top, tip bender with snow-shoe tip locked in place. Bottom, special grooving tool designed to finish bottoms of miners' snow-shoes. Plumas County Museum photos.

In Metcalf's instance, Clinch, who by that time had quit racing, took over as wax man, while "Uncle Bill" continued to run on shoes made by "Old Buckskin" Porter.

Metcalf, a native of Wisconsin, was 18 years old when he arrived in Saw Pit in 1852. It is said that Metcalf participated in the first stake races held during February of 1855. His later snow-shoe career can be followed through old newspaper accounts up until 1870. Thereafter, during the sport's heyday in the seventies and eighties, he played important roles as an official. Metcalf married Eliza Kelly in 1869. She was a young Saw Pit widow and snow-shoeing belle. Their romance was an involvement including two other suitors, Robert Oliver and Bob Francis, who had contested for the "championship of all snow-shoers."

Metcalf won Eliza only after his rivals eliminated themselves by shooting it out in a dance hall brawl. Uncle Bill, as Metcalf was called, died in 1919. He was an imposing six-footer and regarded as "a characterful old gentleman" during his declining years in Oroville.

The early championship stake races were run over courses considerably longer than those used in later years. They were of a point-to-point variety in as straight a line as possible over varied terrain. Such a racing duel was reported in the Feb. 21, 1863 issue of The Downieville Mountain Messenger:

"SNOW-SHOE RACE ON THE BIG HILL TRACK; $100 STAKES"

"During the past week it was given out that William Metcalf, who won snow-shoe laurels at Onion Valley two years ago, was matched with Frank Lee, famous at Whiskey Diggings as a snow-shoe champion, and that they were to run today for $100 a side. Snow-shoe dope, of the chained lightning quality, went up to a high figure immediately. Our Richmond Hill merchant drove a custom house business in the superior kinds of lightning and such coating of shoes and rubbing of shoes and examining of shoes have not been seen on these hills for the past two years at least. The weather this morning was dull, gloomy and disagreeable as it is possible for weather to be.

The storm clouds lay like a thick, heavy mantle over the hills. A chill, blustering, piercing wind drove the light snow before it in streams, and, altogether, it was enough to freeze the marrow of one's bones to be out of doors at all. Still, Poorman's, Hopkins' Forks, Saw Pit Flat, Richmond Hill, Gibsonville and Whiskey Diggings were largely represented at the race grounds at an early hour. The snow was light, heavy, soft, hard, wet and dry—all mixed up together in such a manner that the knowing ones were puzzled to get a coat on their shoes which would run over the track. Many a strong, athletic fellow with the wrong dope had to give in and kick up his heels in a manner very unbecoming to champions and gentlemen.

"There were a few races among the scrubs in the early part of the day for small purses, whiskey, cigars and the like, but on the whole a more quiet crowd of snow-shoe elite has never been assembled in Onion Valley.

"The reputation of Whiskey Diggings and the fame of Onion Valley lay heavily on the minds of all. There seemed to be but comparatively little outside betting until the very last moment. Metcalf fell in a trial trip early in the day, and his friends entertained a good deal of anxiety about his riding his shoes safely. "Little York" Lee, the Poorman's Creek champion, so noisy two years ago, was as quiet and docile as a lamb. The famous "Black Swamp" was as dark and gloomy as the day itself, till three o'clock, p.m. when that gentleman raised his trumpet voice and roared, as loud as a Washoe Canary: "Ten dollars to bet on Bill Metcalf, who will take my ten dollars?" The noise and confusion now became general, and tens and twenties were freely staked on all sides. The betting, however, soon subsided, the noise was hushed and the contestants took their places for the race.

"Both started handsomely, making some half-dozen strokes with their snow-shoe poles in very quick time, that sent them forward with great velocity. Young Lee seemed to be leading his man in gallant style until they came to the most precipitous part of the track, where Metcalf's shoes darted by and led the way to the poles, winning the race by about 15 feet. Time: half a mile in 25 seconds. Allow me to say here, that the time was certain; but as for distance, the deponent sayeth not.

"Many have seen swifter races, but none claim to have seen a more handsome race. Both contestants struggled nobly for the prize, both rode their shoes in splendid style and it seemed almost a pity that either should lose the race.

"Young Lee's friends offered to back him for forty dollars to run over the same ground again, but Metcalf's friends did not seem inclined to take the bet. It is very probable, however, that his race will lead to another at no very distant day."

—EMIGRANT, The Downieville Mountain Messenger.

That winter was a rough one on the reputation of Whiskey Diggings for speedy snow-shoe runners. While Lee was being given a racing come-uppance by Metcalf in Onion Valley, another Whiskey Diggings runner was taking a beating at Howland Flat.

Beer flowed freely among spectators and contestants at "Lost Sierra" snow-shoe races. Above, in 1908 photo, Bill Primeau, left, toasts Champion Joe Bustillos, center at the La Porte championship race award ceremony. The bar, a snow-shoe race fixture, can be seen at right. W.B.B. collection.

And The Downieville Mountain Messenger dutifully reported:

"ALL ABOUT SNOW-SHOE RACES"

"The long expected snow-shoe races; between Jackson of Cold Canyon and all Whiskey Diggings came off last Monday at Sierra Hill, as per program.

"Before I attempt to describe the race proper, I desire to take a casual view, retrospective and otherwise, of snow-shoes as an acknowledged and permanent institution among us. Nearly all will remember the days when locomotion in the winter season was actually impossible except in large companies, or when the snow had become sufficiently firm to support the weight of man; in consequence of which many a poor fellow has passed to his long home for want of grub and a great many others have perished in the snow in the vain endeavor to reach a friendly shelter. Many instances of such deaths must be fresh in the remembrance of nearly everyone. This state of things was completely changed upon the introduction of Norwegian snow-shoes.

William Metcalf

"There is no use denying the fact that snow-shoes are a great institution and in periods of inactivity like the present they afford a very exciting and healthful recreation to the adult male population as well as supply at all proper times a most delightful exercise to the women and children. Nothing could be more innocent, nothing more proper.

"The race before alluded to, after having been postponed on account of weather from Sunday to the first fair day, came off about 2 p.m. on Monday last in presence of quite a crowd of literally fast men from the neighboring towns of Pine Grove, Poker Flat, Cold Canyon, Gibsonville, Whiskey, etc. The Friends of Jackson appeared quite enthusiastic and confident. It was not generally understood until we reached the course—when the fact was announced—that Ike Zent was the man selected by the Whiskey Diggings magnates to battle in their behalf and maintain, if possible, the ancient reputation of Whiskey for fast men. The race, as heretofore announced in The Messenger, was for $100 a side. A great many outside bets were made and probably $1000 was lost and won on the result of the race. I heard of one miner from Poker Flat who invested $300 on Jackson and doubled his capital. A few words will suffice to describe the race. At the word "go" the two men got a fair and even start, on ground comparatively level and before they reached the steep part of the course, where fast running was to be expected, Jackson was so far ahead of his competitor that Zent threw

Eliza Metcalf

out his pole and stopped himself, remarking to his friends:

"It is of no use to run through; this dope will not work." So you will perceive that the race apparently was lost on the part of the Whiskey boys from lack of care in preparing their dope. Jackson, now knowing that Zent had stopped, ran through with, I think, about the average speed of a pigeon. It was estimated by those qualified to judge, that he made the half mile in less than 20 seconds.

"It is our turn to exult (for, of course, we claimed Jackson as living on the south side of Slate Creek) and you may rest assured we do not neglect the opportunity as no one can tell how soon the boot may be on the other leg—dope. At present we have a right to say, "Great is dope and Jackson is its prophet, by whom we all swear—dope."

--WARREN, The Downieville Mountain Messenger.

"Numerous other personal snow-shoe rivalries also were settled that winter. Among these was a match race between two professional men of old St. Louis, namely Henry Fink, the town's cobbler and William Horn, a brewer of "liquid lager" dope, as culled from the columns of the Downieville Mountain Messenger.

"The event made the scene in mid-March. Several hundred residents from nearby camps arrived and stacked their "wooden propellers" in the principal street. There were more than 500 spectators, all with anxious eyes and swiftly beating hearts. They lined the mountainside run to see if the shoemaker's dope could outrun the brewer's fluid magic.

"They're off! Fink, the cobbler, takes the lead. He rides prettily, passing the stakes handsomely, waxing his man by 150 feet.

"Horn weaves unsteadily, perhaps having partaken of too much of his own liquid dope. He falls, a short distance from the finish poles. The purse was $300; length of track 900 feet and the elapsed time, 10 seconds.

"Race followed race to finally end that particular season in St. Louis."

Quicksilver Charley

In his own inimitable style, Quicksilver Charley Hendel reported the last event of the season to The Downieville Mountain Messenger as follows:

"As you are desirous of knowing all facts of local importance, mining news, etc., I will give you an account of the late, and very likely last, snow-shoe race of the season.

"Some time ago, three members of the firm of Wolf and Company had challenged the members of Pocahontas Mining Company, for lager. The race came off last Sunday on a hill near Little Table Rock. A large crowd was assembled to see how Germany would be conquered by Germans and the word was given to start. Jacob Wolf ran against George Gangloff, of the Pocahontas Company, for 10 gallons of lager. The latter passed the stakes about 50 feet ahead of Wolf, his antagonist. The next race should have been between M. Shindler and John Wilsdorf, for 10 gallons of lager, but Shindler, being too much doped at the time, had his place taken by Mr. J. Wolf, who won this race, Mr. Wilsdorf falling when about half way.

"The third race was between Robert Winter and John Shram, for 10 gallons of lager. R. Winter fell when about half way and Shram also fell when near the stakes, but by wading through the snow with one shoe attached, came out victorious. In the meantime 10 gallons of John Wolf's dope, in the shape of lager beer, reached the hill, when almost all present doped their throats to their heart's content. This stuff appeared to be the best for it was all used up in no time. Mr. Ed Danforth then challenged Mr. H. Fink to try the difference of two kinds of dope. Danforth, having the fastest on his shoes, came out a few feet ahead. The multitude dispersed, well pleased, to meet again at John Wolf's to use the other 20 gallons of manufactured dope, which proved, as far as I can learn, too fast for some. Water having become plentiful, all mining companies are at work now."

Folklore of the Lost Sierra includes an exciting legend of how "Judge Lynch" took over following a race in the long, long ago.

While this writer has never found any documentation whatsoever for the legend, it is possible this particular bit of folklore stems from an account published in 1872 by Col. Albert S. Evans, a San Francisco author, in his "Alta California," and attributed to the pen of Col. Charles W. Crocker, a writer for the Oregon Bulletin. Col. Evans in his book pointed out that Col. Crocker had studied "the character and peculiarities" of the early-day miners when they had "peart times on Rabbit Creek."

A condensed version of Col. Crocker's story of March 19, 1852 follows:

"Four long weary months had dragged themselves by since the snow came down upon Rabbit Creek and put an end to all outdoor operations of the miners. For four months the town had been cut off from all communication with its neighbors. The earth was buried beneath the white shroud of snow which had so silently fallen upon it. Four months had elapsed since a mail had been received, and during all of that time the inhabitants of the camp had eaten their food, made snow shoes and waited for news from the outer world.

"A slight thaw, occurring a few days before, had formed a thick crust upon the snow. The residents were now only too glad to visit town, where they could spend a few hours in the drinking saloons or visit the gambling house.

"The gamblers, those who dealt faro, monte and other games of chance, were delighted with the change. For weeks it had been "dog eat dog" with them, and now the prospect of having a few outsiders to fleece was a source of great gratification. In order to celebrate the event they had clubbed together, raised a purse of a thousand dollars, and offered it as a prize to the person who

Snow-shoes, as shown in the above photo of the author, taken at Johnsville in 1952, shows the length and tip style of "racing shoes," used by miners in the "Lost Sierra." Note the broad tip. The shape of tips was the trademark of the snow-shoe manufacturer, and could be readily identified by racers and fans.. Photo by Dr. Frank Howard.

could make the quickest time on snow-shoes over a track to be designated by a committee. The contest was to be free for all who chose to engage in it and it was to witness this race that so many of the hardy sons of toil came into La Porte.

"By eleven o'clock between three and four hundred miners had assembled in the town. All were more or less under the influence of liquor. Around each gambling table could be seen a crowd of hardy fellows betting their hard-earned dust and indulging in rude jests and boisterous laughter. The gamblers were making hay while the sun shone.

"After the noon-day meal had been disposed of, the committee on arrangements set to work to arrange the preliminaries for the snow-shoe race. Judges, timekeepers, referees, starters, etc., were appointed, rules established, and everything fixed in consonance with the ideas of the majority of the committee. Then those who were to take part in the contest were notified to appear at the starting post. The judges took their positions. Those who had been absorbed in gambling left the tables and sought places from whence a good view of the race could be had.

woman outraces man?

"When the hour for starting arrived, the signal was given, and contestants bounded off with the speed of lightning. At the last moment a woman had appeared upon the scene and started with the others. She was evidently an expert in the use of snow-shoes, and passed several of the contestants during the first hundred yards. Those who were watching the race became fearfully excited and whenever the woman would succeed in passing one of the racers, they would make the welkin ring with their shouts of joy and encouragement.

"The race was soon over, and was won by the mysterious female, who had been materially aided by the wind catching in the skirts of her dress.

"Perhaps her success may partially have been caused by the gallantry of the other contestants, who thought it would be ungentlemanly to beat a woman.

"There were but two or three females on Rabbit Creek at the time and consequently great curiosity prevailed to learn which one had entered the lists and carried off the prize, and no sooner had the contestants crossed the home mark than the crowd rushed forward and surrounded them.

"'Who is she,' cried a dozen voices.

"The victor threw back the bonnet and veil that covered and concealed her features, revealing the face of a man, bearded like a pard.

"'Well, I'm danged ef that `ar woman don't turn out to be Jim Wilkinham, who lives on t'other side of the hill,' said Gabe Husker, an honest miner whose curiosity appeared to have been satisfied. "Jim has been playing roots on the boys, and is a thousand

Whipsaw, used to make planks and beams out of logs, installed in a sawpit, to make a pioneer saw mill. Plumas County Museum photo

> "The race was soon over, and was won by the mysterious female, who had been materially aided by the wind catching in the skirts of her dress."

dollars better off fur havin' done so. But dog me ef I don't think the race ought to be run over agin. I wouldn't stand being cheated that way ef I was one of 'em."

"At this moment fierce, angry words were heard within the circle. Several persons appeared to be taking part in the dispute, and again the crowd pressed forward to see what was the matter. Suddenly the sharp report of a pistol rang out, and the crowd fled pell mell. Turning quickly, Husker saw that a murder had been committed. The winner of the purse was lying motionless upon the snow, while the blood, pouring in a stream from a wound in his bosom, was rapidly crimsoning the ground.

"A few feet distant stood Chadwick, proprietor of the town's faro bank, cooly returning his revolver to its scabbard, which hung over his hip. "What in hell have yer been doing," yelled Husker, as he jumped toward the murderer.

"'Bin a givin' a dern skunk his deserts. No dang dead-beat can ever get any of my money by such a fraud upon the community as this one. I go for all sich, every time, you bet!'

"'I guess we'll have to go fur you,' said Husker, as he laid his hand upon the shoulder of the murderer.

"The prisoner, notwithstanding that he was soon bound hand and foot, and entirely at the mercy of his captors, was cool and collected as if he was seated behind his gambling table, shuffling his cards for a lot of greenhorns.

"The question of how the prisoner should be tried was a difficult one to settle. There was no regularly instituted court nearer than Marysville, and to send him there and await the law's delays would occupy too much time and be certain to result in the prisoner's escaping merited punishment. After the subject had been thoroughly canvassed in all its bearings, it was decided to organize a court and have the trial take place immediately. Gabe Husker was chosen judge. Another miner was picked as sheriff. A jury was then selected to try the prisoner, and sworn by the judge to perform their duties to the best of their ability. A person who had witnessed the shooting volunteered to act as prosecuting attorney, and a gambler who had been a friend of the prisoner was sent for to appear and conduct the defense.

"We are away up here in the mountains whar we hain't got no Californy law," Husker set forth, "Therefore, I propose to put it to a vote whether we shall try the prisoner by Lynch Law or Missouri Law. I hold in my hand a copy of the Constitution and By-Laws of the State of Missouri, which are good enough law for me, and ought to be good enough for anyone."

"A tremendous, 'Yes!' went up from the throats of the assembled multitude.

"A number of witnesses testified as to the shooting; in fact, the prisoner himself declared to the jury that he had killed the miner because, "Any dern skunk as would humbug a whole mining camp deserved to have a bullet-hole bored through his diaphragm."

"After the testimony had been taken, the case was summed up in short speeches by the counsel and submitted to the jury. A whispered conversation followed, and then the verdict was announced. The prisoner had been found guilty of murder in the first degree, and sentenced to be hanged by the neck until he was dead.

"The sheriff, accompanied by several men who had been swinging a rope over the limb of a nearby tree, now entered and took charge of the prisoner. The preliminaries were quickly made, the rope placed around the neck of the doomed man, and when everything was in readiness, the prisoner was asked if he had anything to say before he was launched into eternity.

"This 'ere joke has gone fur enough, and as my feet are gettin' cold, I wish you would wind it up. I'm tired of bein' fooled with. I say hold on. I appeal this 'ere case to the Supreme Court of Missouri, and you can't carry out the sentence until the appeal has been decided."

"This change in the aspect of affairs somewhat staggered the crowd, and delayed the execution for a short while. Judge Husker was called upon the give his views upon the subject, and did so as follows:

"The prisoner was tried by Missouri law, found guilty, and sentenced to death by the law; and thar cannot be doubt about his right to appeal to the

Suddenly the sharp report of a pistol rang out, and the crowd fled pell mell. Turning quickly, Husker saw that a murder had been committed.

Supreme Court of Missouri. So fur so good. But courts are always in the habit of goin' on until the Supreme Court issues its mandamus stayin' perceedings. Therefore the sentence of this court will be carried out, unless properly stayed by mandamus. If the preceedin's ain't regular, they can be reviewed when the case reaches the higher court.

"The decision of His Honor was received with a shout, the prisoner said, "All right, go ahead." The sheriff gave the signal and the rope was jerked. The noose broke, letting the murderer drop in the snow. He struggled to his feet and, looking over the crowd, said:

"Thar, didn't I tell you I couldn't be hung? Now, gentlemen, as this show is over, I thank you for you kind attendance, and all of you as has got any money and wants to lay-out at faro, just follow me and I'll give you a lively game."

"He turned to leave the scaffold, when he was met by the sheriff, who held in his hand a much stronger rope. This was soon knotted about the neck of the victim, who looked at the rope and then at the faces surrounding him, but failed to see any sympathy for him.

"He waved his hand as a signal to proceed, and in a moment more the unfortunate man was standing in the presence of Him who judgeth all things."

Out of such factual and legendary incidents as those at Onion Valley, Old St. Louis and La Porte did the snow-shoe stake races evolve into organized tournaments. Intense inter-camp rivalry aroused by the races and betting became too much to be handled by local committees. So a snow-shoe association was organized in 1866 and competitive rules were set up.

References:

Report of a race outside a bar in 1859. Reference: Phil Sinnot, letter cited newspaper account.
1861 William Metcalf, Champion, 1863, Downieville Mountain Messenger.
Alta California, Book at library, Alta California, dated March 19, 1872. Author, Col. Albert S. Evans, Fiction account attributed to Charles W. Crocker of Oregon.

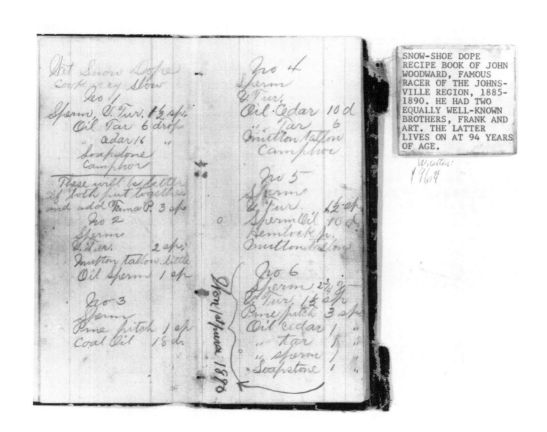

Dope recipe that beat out competition from the Woodward family circa 1890. Dopes were formulated for every conceivable type of snow condition; cold, dry, wet, soft, powder, granular, etc. These recipes were pictured from John Woodward's 1890 dope recipe book. Note first purse winning dope in 1890. Woodward had two equally well-known brothers, Racers Frank and Art. The author met Art, aged 84, in 1969. W.B.B. Collection.

6

The Alturas Snow-shoe Club

The first annual tournament of the Alturas Snow-shoe Club, featuring several hundred dollars in cash prizes and championship belts for men, women and juniors, at stake, was a spectacularly successful affair."
— The author.

The town of La Porte, perched high in a mountain-ringed valley, stands 4,959 feet above sea level and squarely in the center of the Sierra storm belt. This winter scene was shot March 24, 1913, while the town was buried under deep snow. The famed Union Hotel's black roof and row of square windows can be seen at center, backed against the snowy hillside. W.B.B. Collection.

Star emblazoned Alturas Snow-shoe Club racers wait for the starter's gong, left. Snow-shoe racing in the Lost Sierra became a formalized sport in 1867 when organized by the Alturas Snow-shoe Club. It centered at La Porte, a prosperous town surrounded by a cluster of mining communities that all participated in the winter sport. W.B.B. Collection.

At the conclusion of the races a club meeting was held in Legan's Saloon attended by a large concourse of people. It was voted that the next annual tournament of the club, featuring a contest of mountain lightning, should be held at La Porte and commence on Feb. 1, 1868. It would continue for one week and require that three thousand dollars be raised for expenses and purses.

snow-shoe clubs in the racing wheel

Organized snow-shoeing grew "around the "Racing Wheel" of the clubs whose spokes reached out to the cluster of nearby Lost Sierra mining communities. The Alturas Snow-shoe Club was at the hub of the wheel.

The Snow-shoe Era reached new heights during the month of December, 1866, when the preliminary proposal for a formally organized snow-shoe club was set forth in La Porte. The club became a fact in 1867. It was named "Alturas," a name which came out of an unsuccessful drive to create a new county named Alturas.

Earlier that year, after vainly trying to become the seat of a new Alturas County, the town's leading citizens alternatively persuaded the state legislature to move the La Porte region of the Lost Sierra, which today lies in the heart of the Plumas National Forest headquartered in Quincy, from Sierra County and annex it to Plumas County under which jurisdiction it remains.

Alturas means "heights," and was believed to be particularly applicable to the type of organization and the high mountain locale its charter members had in mind.

At that time La Porte, situated 4,959 feet above sea level, was the central trading area for many of the mining communities lying between the main branches of the Feather and Yuba rivers. It rested on a high, relatively inaccessible valley, carved by Rabbit Creek.

In those days it was also the principal snow-shoe racing center for the entire region. Throughout that region many snow-shoeists believed that some kind of a parent organization was needed for the good of the sport, which had enjoyed such a robust and rapid growth. It was to resolve this problem that the club was formed.

The Alturas Snow-shoe Club of La Porte emerged in that manner. It was the world's first club organized for the specific purpose of sponsoring competition and promotional work incidental to such an enterprise.

From Saw Pit, Onion Valley, Port Wine, Gibsonville, St. Louis and Whiskey Diggings, the spokes of the "Racing Wheel," came representatives of the local committees which in previous winters had sponsored races in their towns. Plans were made for a three-day tournament to be staged in La Porte the following February. At that meeting it was also decided to form a permanent organization. Named as interim president was Creed Haymond, attorney, the leading legal light of La Porte, who, in later years, became a national figure in the affairs of the Central Pacific Railroad. To Haymond was delegated the duty of working out the details for the forthcoming tournament.

The first annual tournament of the Alturas Snow-shoe Club was a spectacular success. Racing began Monday, Feb 11, 1867. Several hundred dollars in cash prizes and championship belts for men, women and juniors were awarded.

Due to an intervening blizzard, the tournament was extended until the storm subsided and concluded Feb. 15-16. The weather on the actual racing days was excellent. The lack of new snow on the course during the second day provided an unexpected test for the many racers entered. There was a large crowd, including several hundred visitors from out of town. They packed La Porte's limited housing facilities. All were pleased that everything was systematically conducted. Prizes were bona fide purses and defeated contestants had received fair treatment. Opinions were freely expressed that the successful tournament would go far toward encouraging all future competition.

Creed Haymond, the leading legal light of La Porte. He was the Alturas Snow-shoe Club's first president in 1867. W.B.B. Collection.

At the conclusion of the races a club meeting was held in Legan's Saloon attended by a large concourse of people. It was voted that the next annual tournament of the club, featuring a contest of mountain lightning, should be held at La Porte and commence on Feb. 1, 1868. It would continue for one week and require that three thousand dollars be raised for expenses and purses.

Permanent organization of the club resulted in the election of John Conly, a La Porte banker, to the presidency, while L.D.C. Cray, a local merchant, was named secretary and Alex H. Crew, treasurer. Included among the vice-presidents were Haymond and B. W. Barnes, both of La Porte; Robert Wash of Gibsonville, William

Sturgis of Saw Pit and William Howell of Port Wine.

The miners hoisted Haymond to their shoulders. With manly cheers, they planted him atop Legan's bar where he made a speech in his usual vein of humor. He encouraged athletic sports as a way to develop the miners' physiques and strengthen their social and mental faculties during the long winter months.

There were more cheers for Haymond, cheers for the new champions and last, but not least, cheers for the winner of a leather medal for the "Jackass Championship."

Haymond led three rousing "huzzahs" for the ladies, following which the crowd burst into song, including "No Irish Need Apply" and "Beautiful Star."

The miners' band then struck up a march. A procession was formed and proceeded through a forest of snow-shoes lining the town's main street to the Masonic Hall on the banks of Rabbit Creek. There was a "free-and-easy" dance to conclude the social end of the racing week. The ball in the evening came off without the ordinary five-dollar entry fee. The music played at the race was an essential to the reunion at the hall.

Thus ended what probably was the world's first organized ski tournament. The individual winners of the races therefore may be considered to have become the fast-flying sport's very first United States and World Ski Champions.

early race winners

Crowned as speed king was Robert Oliver, a Cornish miner from Saw Pit Flat, who, therefore, wore the champion's belt for the Alturas Snow-shoe Club, subject to challenge by any snow-shoe expert in the state.

The men's championship, decided in the final day of the tournament, was run over an accurately measured 1230-foot course. Oliver's run was clocked at 14 seconds. There were 42 entrants, divided into six-man squads. The round-robin championship event got under way with George Roubley of Gibsonville winning his heat in 16 seconds and 10 feet ahead of the next best man. Then came William McDougall of Saw Pit, 16 seconds and

Top. Alturas Snow-Shoe Club racers..Bottom, racers of the Alturas Snow-shoe club, March 12, 1910, La Porte. From left, Medev Bustillos, Al Primeau, Clyde O'Rourke, Harold Pike and Will Primeau. 1910 Postcard Photo. W.B.B. Collection.

26 feet; Oliver 16 seconds and 30 feet; Elias Squires of Gibsonville, 15 seconds and five feet; William Metcalf of Saw Pit, 16 seconds and three feet; D. McDowell, Saw Pit, 15 seconds and 50 feet and John G. Pollard, Saw Pit, 16 seconds and two feet.

The finale of the historic speed duel was reported as follows in the Downieville Mountain Messenger:

"The leaders of the different squads then ran for the belt. Robert Oliver and John G. Pollard, both of Saw Pit Flat, made it in 14 seconds. Being a tie, these two gentlemen again ran, Robert Oliver then coming out four feet ahead—time, 14 seconds."

Oliver's winning time of 14 seconds for 1230 feet, even from a standing start, was not considered very good in those days. Before the Snow-shoe Era ended the speed made by Oliver of 60mph in the first championship was to be exceeded many times and had reached 90 miles per hour.

Quotes from newspapers in the region give the atmosphere and excitement of snow-shoe racing:

"Horse-racing is as nothing compared with the excitement of snow-shoe racing. Of course, the time made by horse or railroad racing is not to be compared with the time made by a snow-shoe racer. Only the bird of prey swooping down from the lofty summits into the mountain canyons, can compare in time, with the speed during the races just reported.

"Long ropes were stretched up and down the hill, about 100 feet apart. This was the track, with poles fixed at the upper and lower ends. As it would be impossible for all who entered the race to run at once, they ran in squads from five to seven, the winner of each squad standing aside to come in at the final "saw off" race. They were placed side by side at the starting place, and went at the gong as a saw blade was hit or the tap of the drum. Then down they would go as if shot from a cannon; scooting and sliding, first one ahead and then the other. Now and then an unlucky hombre would lose his balance and plow through the snow, reminding one of the antics of a straw man thrown from the top of the hill by a strong arm.

"On Monday evening a complimentary dance was given to visitors by the citizens. On Wednesday evening the Masonic fraternity gave a complimentary dance to visitors; and on Thursday and Saturday evenings complimentary dances were repeated by the citizens for the

John Conly, La Porte banker, 1868 president, Alturas Snow-shoe Club. Jann Garvis Collection.

Robert Oliver, speed king, winner of the Alturas Snow-Shoe Club's first Championship meet at La Porte. W.B.B. Collection.

visitors.

"The attendance at the dances was large—Everything was lovely and the goose hung high."

Because complete results of the Alturas Club's first tourney are unknown to modern-day skiers and have not been re-published since the early days of the Snow-shoe Era, the author believes it pertinent to include them here.

Miss Lottie Joy won the women's championship in 17 seconds and finished ahead of her nearest competitor by 10 feet. An early-day contemporary reporter wrote that Lottie was, "the snow-shoe pet of St. Louis. She dropped low, with the pole under her arm, and just scooted down the track like an arrow to the mark, while the others, carrying too much sail, with shoes wide apart, came through all standing up, but too late to win."

Winning the junior championships were Henry Kelly of Port Wine and Miss Hattie Starr of Gibsonville.

Six other purses, in addition to the championship event, were contested during the three days of racing. The individual winners were Frank Saulter, Gibsonville; Robert Oliver, Saw Pit; George Roubley, Gibsonville; E. G. Squires, Gibsonville; Napoleon Norman, Saw Pit and Peter Riendeau, Saw Pit.

It is interesting to note here the names of Robert Oliver and Robert Francis, both of Saw Pit, as having been among the squad winners in the tournament's very first race, a contest for a $40 purse won by Peter Riendeau. Although Oliver and Francis both came from Saw Pit, they

Ladies of the Lost Sierra dressed up for many occasions, but never went anywhere in winter without their snow-shoes. Above, five young matrons posed for a picture, holding a pair of snow-shoes at La Porte, 1908, to show their length. W.B.B. Collection.

Spectators gathered at the race courses to cheer their champions and watch them perform. For seated spectators, pine boughs were cut to make things comfortable. The ladies came to the race course on snow-shoes and parked them nearby. W.B.B. Collection.

were bitter personal rivals. Out of that rivalry was to come one of the most hotly contested criminal trials in Plumas County history following the killing of the world's first snow-shoe champion by his fellow townsman in a saloon brawl.

However, the killing of Oliver by Francis and the latter's subsequent trial could not be foreseen in February of 1867 and meanwhile the reader can follow other developments that took place around the snow-shoe racing circuit.

The Alturas Club's tournament had opened the racing season and other mining communities now prepared to follow suit. At Port Wine a snow-shoe club was organized, while other snow-shoe racing centers continued with the old-fashioned committees in charge of the events. There had been no serious accidents during the races in La Porte, but the riders were not so fortunate once they took to the snow-shoe racing "wheel."

From Port Wine came the following account from a snow-shoeing correspondent:

"There was one sad occurrence to mar the unbounded happiness of all. Napoleon Norman, winner of a prize at La Porte, having won his squad, was a contestant for the prize. On the way down the track, he stumbled and fell over a cavity in the snow where another man had fallen. He dislocated his right leg at the thigh joint, and fractured the bone of the joint, besides receiving other internal injuries caused by the fall. The purses of the winner were made into one, other money was added to it by citizens, and the whole amount presented to the unhowever, was that the racer broke one of his shoes caught in one of the lady's hoops, without serious injury to either party.

Actually, however, the excitement at Port Wine was just getting a good headway. The next day Miss Lillie Brown had her shoulder badly hurt in a fall. Miss Mary Lloyd was struck on the shoulder by a runaway snowshoe and was run over by another lady. To make matters

Members of the Port Wine Snow-shoe Club, gathered at La Porte in 1897 for the above photo. Some racers, as the one at center, featured skis over 12 feet long for an attempt at extra speed and stability. W.B.B. Collection.

fortunate man."

Charles Barnes of La Porte was the winner of the first day's principal race at Port Wine, covering a track 1025 feet long in 12 seconds. The final squads for the secondary purse were enlivened when one of the racers, while going at about the "slow speed of greased lightning," collided with a young lady, throwing her, "Canada West," and himself, too. The most singular aspect of the incident,

worse, Miss Lottie Joy of St. Louis, the newly crowned champion, refused to run the final squad for the money.

Officials had decided the course was too dangerous and shortened it to a lower starting point. Lottie, the pride and joy of the newspaper reporters, insisted that she would run any track the men could, and when her claim was disallowed, she defaulted to Miss Lloyd.

The men's races continued and several riders required medical attention. Charles Clark had one of his knees unjointed, while Mike Kelly's nose was badly disfigured by a splinter from a snow-shoe running through it. Both men were Port Winers and had their boards doped with a concoction know as Kleckner's "No. 1 Plain and 4 O.T.", which was lightning.

Snow-shoe racing was a dangerous sport. While more timid officials preferred to separate the racers anywhere from 10 to 15 feet, riders for the big money often ran almost shoulder to shoulder in four, six or eight-man squads. It is easy to determine what could happen to them when a middle man spilled as they shot down the mountainside.

The 1867 racing season reached a peak at Howland Flat, March 11, when "seven hundred people, of whom quite a proportion were ladies, lined the Table Rock course for the opening events of a four-day program."

Out of that tournament came the organization of the Table Rock Snow-shoe Club. Its membership came from Howland Flat, Poker Flat, Pine Grove and St. Louis. Its first specific commendation was the ability of its three youthful riders, whose names for many years became household words in the Lost Sierra.

"Among those deserving special mention for expert

Race spectators are shown arriving at the La Porte Hotel in this 1890 photo. They came on snow-shoes and sleighs. The horses, shown above, were wearing snowshoes, which were commonplace at that time, as they were being used for regular mail delivery. W.B.B. Collection.

riding," a newspaper writer reported, "I may mention Elias Squires of Gibsonville, Charley Littick of Pine Grove and Yank Brown of Howland Flat."

It was Sam Wheeler, a Howland Flat storekeeper, who was responsible for founding the Table Rock Snow-shoe Club. For years he had acted as local racing committee secretary, but believed that a "greater unity of action was needed."

D.H. Osborne of St. Louis was named president and Wheeler continued as secretary. Out of the first meeting came the announcement, "to offer a large number of prizes to be run next winter, and with the Alturas Club pursuing the same course, we may look for better results and more extended proficiency in the dope."

The winter of 1868 was one of monumental snows. Tournament racing opened Feb. 3, with a five-day program in La Porte. It continued each week throughout the mountain communities until the season's final events, April 10, at Port Wine.

Six hundred dollars in cash and $300 worth of silver watches, in addition to the usual entrance fees, were at stake. Weather conditions were good and the track "beautiful," but snow conditions must have posed a tricky waxing problem because racing times were comparatively slow. The fastest time over a 1400-foot course was 21 seconds by Peter Riendeau. The Saw Pit rider won not only the championship belt, but $325 in cash as well. For the major event there were 72 entrants, Riendeau outrunning such stalwarts as Robert Oliver, the defending champion, Pell Tullo, of Grass Flat; Bonney Abbott of Port Wine, Robert and William Francis of Saw Pit; Charley Scott of Poker Flat and George "Yank" Brown of Howland Flat, a teenager contesting in a championship race for the first time.

Frenchy Riendeau continued his winning ways around the racing circuit that year, closely pressed however, in the amount of stakes won, by Robert Oliver and the up-and-coming young Yank Brown. Excitement and interest increased as the racing season progressed. At one time The Downieville Mountain Messenger issued an extra edition in order to give its readers a comprehensive picture of doings at Howland Flat, where the first tourney of the Table Rock Snow-shoe Club was held in March. At the same time The Messenger editor gloated that Benny Jones, one of his snow-shoeing newspaper carriers on the route between Eureka and Port Wine, had won a race at Morristown and was "hard to beat on snow-shoes by anyone his weight and size."

On April 8 1868, as the second season of organized racing was ending, a Port Wine correspondent, signing himself, "Snow-shoeist," sent the following essay to The Messenger:

Henry Sibbly, winner of the first day's race at La Porte, Alturas Snow-shoe club race, March 23, 1894. W.B.B. Collection.

"RIDING SNOW-SHOES"

"I will now give you my idea of riding the shoes, which will embody the experience of more than nine years incessant practice upon them as regularly as the winter rolled around.

"Firstly, I would say to those who are about to make their first attempt at riding shoes, that the one great requisite to possess is confidence—they must at least feel a fancied security.

"Timidity will do in no instance, and for one, on starting down a hill, to be afraid of falling will never do; he might with as much success try to stem the current of the Niagara River as to keep from falling when he thinks he may, or has not the confidence within himself. What is needed in a beginner is conceit, though it will not do in all cases for him to possess too much, as it may lead him into danger; but a little of this self-important element, without foolhardiness, is essential to give one courage and smooth the way to success.

Many have learned to ride shoes in an incredibly short space of time, and made fair rides with but very little practice.

"To beginners I would suggest as to the best manner of riding and managing the shoes, or the ways which I have found the best, that is to ride very low upon the shoes, or in what is called by snow-shoeists the squatting position, and to hold the pole in the right hand, I find the best suited to myself; and in going over a jump or a raise in the snow, which is occasioned at times by a tree lying across the track, or by the wind drifting and forming a hollow and raise, which will, when a snow-shoeist is going fast, make considerable lift of both shoes and rider, and sometimes the shoes go on their course alone, while the rider is making strange gyratory motions in the air, a thing not uncommon with beginners upon these uncertain carriers—and here I speak from experience in an extended practice—the best plan to ensure success is to raise partly up, at the same time gently waving the pole and left hand.

"This will greatly ease the rider over such places, and when fairly over he can easily resume his original position, thereby gaining speed.

"There are great advantages in the present style or riding over that which was in vogue several years ago, and was known as the standing erect position. The body in passing through the air quickly would naturally encounter more or less wind and be somewhat impeded. The late style has entirely superceded

Fun-filled miners gathered at La Porte for this 1897 photo taken in front of bar #2 which provided plenty of "liquid dope.". A snow-shoe entrant, with stars on his shirt, frolics in the foreground. W.B.B. Collection.

the old, as the resistance of the wind, is to a great extent, surmounted by the low position; and by leaning well back upon the shoes and riding, as it is called, upon the heels, the speed is accelerated.

"In racing it requires considerable skill to get a good start, and to secure this the pole is used as a propelling power, and he who best exercises it stands the best chance to win—provided his dope is good. In conclusion I would say to the uninitiated in the mysteries of snow-shoeing who have a desire to learn the same, that with confidence all may command success."

For the 1869 season, great plans were made, and the fraternal organizations, which always played a large part in the community life of the early mining camps, were out in full force and regalia on snow-shoes. From Sierra City the ancient order of E. Clampus Vitus crossed the mountains in a body and it was "Hail to the Clampers," in Howland Flat. Downieville's Cosmopolitan Minstrels snow-shoed to Gibsonville, while from Pilot Peak the Odd Fellows slid to join their brothers in La Porte.

Lively scenes were enacted in all the racing centers, with many tall, trim-built racers looking defiantly at their competitors as, with sly winks and nods, sacks of dope were unstrapped. Local bands tootled in the snow, saloon-keepers moved their bars to the race tracks, and there were night-long dances and much entertaining in private and public.

The Lost Sierra now was the snow-shoeing center of the world, with like activities on an organized basis still some 10 years away in Norway, the motherland of the longboards.

In the news columns of The Mountain Messenger a correspondent inveighed against Dick Alley:

"...' a hitherto respected citizen of Howland Flat.' With much sorrow and deep chagrin' it was reported that Alley had 'put up a job on the snow-shoeists' the previous season by having bought heavily in the pools, always taking third choice, he mixed some lightning dope, warranted to run even onto the electric current; to tell the truth it did run well, for it was of gratuitous distribution and that is where the fraud comes in. No one could ride it, and the user was a good bet, 2 to 1, to fall. The more dope one used the less chance he had of winning his squad. Moral: beware of Alley's liquid dope."

Sondre Norheim, known as the father of modern skiing, has been credited with staging the first organized ski races in Norway. The author's documentation indicates that the Alturas Snow-shoe Club came first.

The 1869 season opened with the Port Wine Snow-shoe Club's tournament. From all the mining camps hundreds converged on Port Wine. Their eight to ten-foot traveling shoes cut tracks across ridges and down canyons.

All had packs on their backs, while the men and women intent on entering the speed events were doubly burdened by carefully wrapped long racing boards carried on their shoulders.

Immediately, the topic of the times, paying mines sank into insignificance when compared to the all-absorbing question of dope, which was on the minds of all and was bound to stick there until the racing season was completed. The races began over an icy track, with Frenchy Riendeau winning the opening event. Many of the Port Winers had lightning dope, but were afraid to ride it, several going down on their poles. The local boys, however, brought their dope to bear the next day, when Bill Robinson eased Frenchy out of the big money.

From Port Wine the racing crowd wheeled over to La Porte, where what since has become known as the first "international tournament" opened for a five-day stand. On hand to compete and, if possible, take home with him to Alpine County the many hundreds of dollars in cash purses was none other than John A. "Snow-shoe" Thompson, the redoubtable Norwegian express man whose feats in those years were widely heralded throughout the mountains. Thompson, then 42, was unable to compete successfully against the youthful dope riders; and went down to defeat in the very first squad race.

Racking together $250 in purses and the championship belt was Yank Brown, with Frenchy Riendeau a close second, followed by Charley Littick and Orville Squires.

The following week at Howland Flat the championship went on the block again, with Yank proving too fast for all his challengers.

Moving across the canyon to Gibsonville, the racers contested a series of six club purses, Yank winning one and Littick, two.

The social end of the Gibsonville races was reported by "Michigan," who had this to say:

"In the evening an entertainment was given by the Gibsonville Dramatic Club. Several pieces were very credibly performed for amateurs, and pleasantly interluded with songs and fine music by our well-practiced band. Sheridan's Ride to Winchester was beautifully rehearsed by Charley Littick. Dancing kept up into the morning hours...The snow is deep, covering many of our cabins, all save the stovepipes which stick out a few feet above the snow. Yet our town is not so still and lonesome a place as it might be. The boys are lively and full of sport."

Among the liveliest that winter was a youngster named Tommy Todd, whose name appeared in the newspapers for the first time following the Howland Flat tournament, where he won the junior race. This youth was to reach his majority five years later and then set an all-time speed record.

To appreciate the old-timers' fast-dope fraternity one has best to compare other varieties of speed with that of snow-shoeing.

The velocity of a race horse is from 29 to 33 miles per hour; of a cloud in a violent hurricane, 80 to 100 miles, but not even a jet exceeding the speed of sound can faze the true Lost Sierran.

In the autumn of 1875, the two leading dopemakers; of Howland Flat, Hiram Walker and A. J. Howe, reached a momentous decision. They believed greater speeds could be obtained, not with improved lightning dopes, but with better-designed snow-shoes. It now becomes pertinent to point up the two men's backgrounds. Howe, who later became a Sierra County district court judge, was born in Attica, N.Y., and had come to the mining camps in 1853, and at the time he joined ideas with Walker, he was a counselor in old St. Louis. Walker was a Howland Flat mining engineer and later was to become better known as superintendent of the Bald Mountain Mine near Forest City in the Alleghany region of Sierra County. Both men had much ability to apply to their project.

Working in great secrecy the two collaborators designed a pair of snow-shoes that would go down in racing history. A slightly wider groove was tooled out, the ends of the long spruce boards tapered round and the tops were ridged and streamlined as on the better modern skis.

Fitted to Tommy Todd, their bottoms well polished with lightning, the new boards did the trick. During the annual races at La Porte in March, 1874, a new record, startling even to those used to mountain speeds, was set when Todd shot down an 1804-foot-long course in 14 second flat from a standing start.

That was at the rate of almost 88 miles per hour for the entire distance covered. When it is considered that for the first two or three hundred feet, Todd's velocity must have been considerably less, only lightning can be compared to the streak he must have been at the finish poles.

The snow-shoeing champions were men like Robert Oliver, a Cornishman; Peter Riendeau, a Frenchman; George Brown, a Yankee and Tommy Todd, the fastest of them all.

Few, if any, Scandinavian names graced the champions' roster, once known to all who lived in the Lost Sierra.

References:

Stories from newspaper accounts in the La Porte Union, and the Downieville Mountain Messenger.

This advertising and program was run in the Downieville Mountain Messenger. Sutter's Fort Historical Museum.

7

The Man From Silver Mountain

"He flew down the mountainside. He did not ride astride his pole or drag it to one side as was the practice of other snow-shoers, but held it horizontally before him after the manner of a tightrope walker. His appearance was graceful, swaying his balance pole to one side and the other in the manner that a soaring eagle dips its wings.
—DAN DE QUILLE, VIRGINIA CITY TERRITORIAL ENTERPRISE.

Snow-shoe Thompson, shown as the artist conceived him, braking with his pole. Snow-shoe usually rode his shoes with his balance pole held horizontally in front of him, poised like a tightrope walker as he soared down the mountain side as though he were a ballet dancer. Woodcut from Hutchings' California Magazine, Vol. 1. No. VIII, February 1857.

John A. (Snowshoe) Thompson, a few days after his death, May 15, 1876, was credited by the Virginia City Territorial Enterprise for his winter assay trip across the Sierra Nevada eighteen years before that brought world attention to the big gold and silver strike on the Comstock:

TERRITORIAL ENTERPRISE
May 19, 1876

"Away back, eighteen years ago, Snow-shoe Thompson carried over the Sierra from Genoa—done up in the remnant of a check shirt——some strange black material, which was bothering the miners who were working in Gold Canyon, about where Gold Hill is now, or perhaps below there. He showed the "stuff" to Professor Frank Stewart, who was then managing a newspaper in Placerville, and asked him what it was. Stewart said instantly: 'Sulphuret of silver mixed heavily with gold.' Thompson carried the material to Sacramento and had it assayed, and the result was several hundred dollars per ton in gold and several hundred dollars per ton also in silver.

"But for that little circumstance it might have been years before the wonders which have been going on here for seventeen years would have been started."

"Snow-shoe" — few ever knew that his right name was John A. — was born in the Telemarken district of Norway, April 30, 1827.

He was baptized and given the name: John Tostensen. John, as a very young child, accompanied his father on trips to other villages over the Norwegian mountain. He became expert in the use of "snow skates," as they were called. They were long, slender blades of wood, curved at the front and ending in a point with the foot resting just a little back of center and held in place by a single strap over the toe.

When he was 10 years old, his father having died, his mother brought him, with his brother, to America. They were accompanied by a friend from Kongsberg, a city which gave its name to the town in Alpine County, California, which later became Snow-shoe Thompson's home town of Silver Mountain.

The Thompson family joined a group of 50 farmers from Tinns. In 1838 they left Telemarken on the long journey to America.

First the family settled in Illinois, then Missouri, then Iowa and back again to Illinois. In 1851, then 24, the man who was to become known as Snow-shoe joined a party headed for California's gold fields. He mined for a time in several camps and later, when success did not crown his efforts, turned to agriculture.

In California's Kelsey Diggings he met a pink-cheeked English lassie, Agnes Singleton, who had come to California with her mother and brother. The latter were not so

Snow-shoe Thompson born April 30, 1827, died May 15, 1876, shown in his later years as a pillar of Alpine County. His portrait now hangs in the Norwegian Embassy, Washington D.C.

sure they approved of the match between the striking blond, blue-eyed Norwegian and the dainty English miss. The couple, nevertheless, soon married. The ceremony was held at Empire, 30 miles from Carson City in May 1866. They established their first home on a small ranch on Putah Creek.

Farming had no more appeal to Snow-shoe, however, than the mines had. In Sacramento newspapers he read about miners' difficulties during winter time and, as a result, his eyes and thoughts swung toward the distant snowline and white peaks of the high Sierra.

One has only to recall that the discovery of gold had brought a migration of 150,000 men into the northern portion of California in the years from 1849 to 1851—and other thousands followed. Their most important link to the homeland was the letters which they could send and receive.

The quaint Lutheran church a in this 1976 photo, was attended by Snow-Shoe Thompson before his departure for America in 1837. The church is close to Tinnsjo Lake and below the Rue farm where he was born. W.B.B. photo.

An increased demand for communication between Eldorado and the East had resulted in the establishment of mail routes between San Francisco and Salt Lake City in 1851, when the chore was contracted for by George Chorpenning and Absolom Woodward for a $14,000 annual stipend.

A monthly trip each way was to be made. Chorpenning took the first mail out. He and his man left Sacramento, May 1, 1951, to follow the "then traveled trail" which led a distance of 910 miles across the Sierra and on toward the shores of the city by the Great Salt Lake.

For 16 days, with their mail and supplies packed on horses, the party struggled through snowbound mountain passes. They came down the eastern slope by way of what is now called Hope Valley. Wooden mauls were used to beat down snow and break trail for the animals while crossing the high Sierra. The party was exhausted when it reached the Carson Valley, where it rested for several days near the site of the Mormon station, established that July. It later bore the name Genoa when the Utah Territory became the State of Nevada.

Chorpenning and his men were 36 days, all told, completing the first trip through to Salt Lake. He and Woodward continued to carry the mail by this route during the following months of 1851. The mail contracting firm purchased supplies at Mormon Station in Carson Valley when its protective stockade was completed.

Hostility of the Indians added to the difficulties of mail transportation. They killed Woodward in November, 1851.

During the winter months that followed several attempts were made to cross the mountains in December and January. The men and animals were forced back by deep snows and furious storms. The February mail from California had to be routed up the Feather River Canyon and over Beckwourth Pass to connect with the trail following the Truckee and Humboldt Rivers. This developed into a harrowing 60-day trip for the mail carriers. It became so cold on the Nevada desert that their horses froze to death. The men had to pack the mail and what supplies they needed on their backs the remaining 200 miles to Salt Lake. During the summer of 1852 the mail was carried through regularly, but never again did the firm take it through the full distance in winter. Instead it was sent by steamer to Los Angeles, from where it was routed eastward over the Old Spanish Trail through Arizona.

This change in the mail route meant that, unless some other means could be found, the residents of Carson Valley were to have absolutely no communication with the outside world during the several months when the mountains were snowbound. It was then that two men, Fred Bishop and another, named Britt, took over the task of linking the western and eastern slopes of the mountain country. To travel over the snows they used webbed snow-shoes of the Canadian or Indian style. Soon they were joined by George Pierce, who also tramped on basket-weave style snow carriers. Little is known of these men and by whom they were paid, but it can be presumed that they charged a set sum for each item of mail delivered and so much per pound for express.

The name of John A. "Snow-shoe" Thompson was added to the roster in January, 1856, when the Norwegian farmer from Putah Creek, California, made his first successful crossing from Placerville to Genoa, skiing from the snowline at Strawberry, California, to the snowline at Woodsfords, California, a short distance from the Nevada state line and Genoa. Soon he was the only man who would and could stand the mountain furies and he took over the Siberia of snow on that route for himself, leaving the shorter and less arduous ones to other men who were not so daring.

Thompson's Viking spirit had been aroused by newspaper accounts of mail carriers' difficulties. As a youngster he had learned how to handle strips of wood attached to his feet. California folklore does not place Thompson on snow-shoes prior to December of 1855, but recent evidence from old diaries in the Placerville region, reported in "Notes on History of Snow-Shoe Thompson," by former State Sen. Swift Berry of California, credits a Norwegian with using snow-shoes for mail deliveries during the winter of 1852-53. That unidentifiable man well could have been Thompson, who had been at nearby Kelsey Diggings and Coon Hollow throughout that winter.

In any event, Thompson did call upon his memory to make himself a pair of snow-shoes. They are said to have been of oak, weighing 25 pounds. Needless to say, they were no burden to the man who was to trust himself to the terrors of the mountains on such fearsome boards. He was as strong as an ox.

The Rue farmhouse at Tinns, Norway, showing the outbuilding, at left, where Snow-shoe Thompson spent his childhood with his parents Tosten Olsen and Gro Johnsdotter.

Thompson is said to have practiced in secret during December of 1855 in the snows above Placerville. The great day upon which he was to display his prowess in public soon arrived. His friends did not share his enthusiasm. They begged him to give up the idea of carrying mail across the mountains on such contraptions, swearing he would dash his brains out against a tree or plunge down some canyon to break his neck. But Thompson chortled at their fears and, with his feet firmly braced and wielding a heavy balance pole in his hands, he flew down a mountainside. He did not ride astride his pole or drag it to one side as was the practice of other snow-shoers.

"He held it horizontally before him, after the manner of a tightrope walker. His appearance was graceful, swaying his balance pole to one side and the the other in the manner that a soaring eagle dips it wings," wrote Dan de Quille of Snow-shoe Thompson in the February, 13, 1876 issue of the Territorial Enterprise.

Thompson carried the trans-Sierra mail all during the winters of 1856 and 1857. Of him, during that period, it has been written that he was under a $200 monthly contract with T. J. Matteson of Murphy's Camp, and that he had continued to carry the mail without a contract for many more years to come, according to former State Sen. Swift Berry's history.

The route from Placerville to Carson Valley stretches 90 miles, much of it snow and ice-bound wastes. Throughout the crossing Thompson soon had cached extra pairs of snow-shoes when he found that the snowline varied with the severity of storms. On most of his trips his pack weighed between 60 and 80 pounds. It was not all mail. Often there were medicines, ore samples and knick-knacks included, as he also acted as a purchaser and deliverer of hard-to-get items for residents of both the mountain country and Carson Valley.

His eastward trips were the most difficult as he faced numerous steep climbs. He usually required three days to reach Genoa. He never took longer than two days to complete the westward crossing. History credits him with being able to cover the distance in much less time than that.

At first Thompson provided necessities for the isolated residents of the Sierra's eastern slopes. It was there that the pioneers engaged in agricultural pursuits. For the most part they were desperately poor. He never failed them during the winters of 1856, 1857 and 1858 while acting as a mail carrier.

During the summer of 1866 Thompson, who in 1858 had homesteaded land in Diamond Valley, 30 miles south of Carson City in Alpine County, established a home there on the eastern slope and brought his wife across the mountains. He was a born naturalist and named many of the surrounding peaks and valleys. A finely watered defile, holding promise as a future settlement, he called "Hope Valley." An adjoining one, not so inviting, he called "Faith Valley," while a rocky, uninviting valley he christened "Charity." And so they are called to this day, according to the Thompson biography written and published by Snow-shoe Thompson Chapter NO. 1827, E. Clampus Vitus.

During the summer and autumn of 1858 a wagon road was constructed across the Sierra. El Dorado and Sacramento counties paid for most of the cost. Then came the overland stage line, in which Thompson and a "Judge John A. Childs, of Genoa," operated a section through the upper most elevations and over which they introduced horse-drawn sleighs during the winter of 1859.

Next came the Pony Express, the Comstock Silver Era, new stage lines, thousands of teams and, finally, across the central route over Donner Pass came the Central Pacific Railroad.

Meanwhile Thompson had always been ready to travel about the Sierra. He explored isolated regions which required supply-packing, possible only by a man with prowess on snow-shoes. If through progress, he ever experienced any regret at the prospect of his business being ruined, he must have soon drawn some consolation from the reflection that his back would not have been strong enough nor his snow-shoes fleet enough to have fully satisfied the heavy demands of the public after the mines of Nevada's Comstock Lode began attracting thousands which came streaming across the mountain passes.

That Thompson's trips across the Sierra, during the three winters he is known to have actually carried the mails, were extremely hazardous, cannot be doubted— and that he was paid mostly during those years with promises is a part of the record. Snow piled to great depths in the Echo Summit region between Placerville and Lake Tahoe on the route to Carson Valley. Through pine forests by day he followed a blazed trail, by night he had only the stars and his natural instincts to guide him. Frequently he was enveloped by storms. One particular place, Cottage Rock, served as a rest station. There he had a small cavern, no bigger than a baker's oven. In front of this he would light a fire and then crawl inside to wait out the storm. On other occasions the snow-shoeing mountaineer danced Norwegian halings on some boulder as a storm swept past in order to keep from freezing. He seldom carried either blanket or overcoat, because the normal temperatures in the high Sierra do not usually stagger down to bewildering zero depths. His provisions were dried beef and crackers, tucked away in the bosom of his flannel shirt. Many moonlight nights he made a bed of pine boughs and camped on the field of snow. Only the fiercest storms ever barred his passage.

Others might become lost, but not Snow-shoe Thompson.

"I was never lost. I can't be lost," he once asserted. "I can go anywhere in the mountains, day or night, storm or shine."

He was wont to tap his forehead with a forefinger and repeat:

"I can't be lost. I've got something here that keeps me right. I have found dozens of men, first and last, but I have never been lost myself. There is no danger of getting lost in a narrow range of mountains like the Sierra, if a man has his wits about him," he told Dan deQuille of the Territorial Enterprise.

Thompson rescued many persons from isolated areas in the early days of his self-appointed tasks. The first such incident, and one of the more notable, occurred during Christmas week of 1856 when he saved the life of James Sisson, a prospector who had become snowbound in Lake Valley. Sisson was in a pitiable condition when discovered by Thompson. For 12 days he had been lying in a lean-to with both legs frozen, and with only raw flour as a means of sustenance.

Thompson rendered what first aid he could and, after first making Sisson comfortable, proceeded to Genoa. There he persuaded six men to join him in an effort to return and rescue Sisson. Late that night they reached Sisson and, using snow-shoes, contrived a crude sledge. The next morning they began the return trip, but two feet of new snow had fallen and it was not until the following day that they reached Genoa. Here the settlement's only doctor decided that the only thing which possibly could save Sisson was the amputation of both his frozen extremities. He declined to perform the operation without administering an anesthetic.

Ether was to be obtained only in Sacramento, 150 miles away. Thompson started back over the trail and returned with the precious fluid in time for the operation to be successfully performed.

Much had been printed in the newspapers about Thompson prior to his celebrated racing controversy with the Plumas and Sierra boys. He was a particular favorite of editors on both slopes of the Sierra in his own particular domain. To those on the western side he brought news from over the mountains; to those on the eastern slope he brought badly needed material as well as news. It was Thompson who, in such a manner, materially aided in the establishment of the Territorial Enterprise, which later, at Virginia City was to blossom into national fame with such writers as Mark Twain and Dan de Quille. The paper's first edition as a weekly sheet was issued on Dec. 15, 1858, at Genoa, using type faces and material packed in by Thompson, who carried the freight through use of both snow-shoes and pack animals.

Thompson's mail-carrying for three winters and his later self-appointed tasks as a messenger of mercy were all more or less a labor of love. He was paid on a niggardly scale for mail-carrying, and for many years had believed he was entitled to greater recompense. There had been many promises made to him on both sides of the Sierra, but little cash other than tips ever evolved. It is said that during the summer of 1868 the newer and more well-to-do residents of Carson Valley drew up a petition to be presented to the Nevada Legislature in hopes that Thompson, at that late date, could in some manner be rewarded for his efforts in delivering the mails and aiding the early settlers. Be that as it may, it is apparent at this late

Silver Mountain City in Alpine County, is now a ghost town. Once the home of Snow-shoe Thompson, it was originally called Kongsberg after the city near Norway's silver mines. Kongsberg is a skiing mecca which produced such outstanding U.S. skiers as the late Roy Mikkelsen, Auburn, CA, 1932-1936 U.S. Olympian, and 1953 and 1955 National Ski Association jumping champion.

Another Norwegian, Roy Mikkelsen, famed member of the Auburn Ski Cub, twice U.S. Ski Jumping Champion, and former Mayor of Auburn, was a native of Kongsberg, Norway.

day that the people of Nevada then knew Thompson had some kind of debt owing him, even though it was not collectible in his own immediate neighborhood.

On Feb. 20, 1869, while Thompson was absent in the Plumas and Sierra snow-shoe country, the Nevada Legislature passed a resolution urging the Congressman from Nevada to work in Congress for $6,000 compensation for Thompson for carrying the mails in 1856, 1857 and 1858.

This resolution was presented to the United States Senate, Jan. 19, 1871, by Sen. William M. Stewart. It was then referred to a committee, which later recommended the passage of a scaled-down claim.

The committee report, a concise snow-shoe saga, from the Congressional Record follows:

" During the winters of 1856 and 1857, while the people residing east of the Sierra Nevada Mountains were cut off from all communication with California, the claimant, J. A. Thompson, undertook to, and did carry the mails from Placerville to Carson Valley, a distance of 90 miles. Twice a

A Diamond Valley monument near the ranch where Snow-shoe Thompson lived until his death, was dedicated in 1958 by Calif. State Sen. Swift Berry, right, while a fellow member of Snow-shoe Thompson Chapter No. 1827 E. Clampus Vitus, blew the ceremonial hewgag.

A statue of Snow-shoe Thompson was dedicated at the Western American SkiSport Museum, Boreal Ridge, I-80, California, May 18, 1976, 100 years after his first mail carry across the Sierra Nevada. Attending from left, author Bill Berry, Mac McCabe of the Snow-shoe Thompson Chapter, Sons of Norway and Sonen Christian Sommerfelt, Norwegian Ambassador to the U. S.

month with regularity he performed the perilous journey over the Sierra Nevada Mountains, traveling on snow-shoes, with a mail sack averaging 40 pounds on his back. For a distance of 60 miles there was no habitation on the way, and all traces of the road were obliterated by snows from 15-20 feet deep; and all travel obstructed from five to six months each year. No man save Thompson could be found who was willing, under the circumstances, to transport the mails across the mountains; and he was induced to do this only by the earnest solicitations of the people of Carson Valley, and the hope that in the future he might be paid for his services.

"On the First of January, 1867, the postmaster at Carson Valley wrote to the Post Office Department that the lowest sum for which that office could be supplied with mail, monthly or semi-monthly, from Placerville, was $1,000, and that J. A. Thompson was willing to carry it for that amount; and recommended him as the proper man to perform the service. At the same time he stated that Thompson was then carrying the mail, and charging $1 per letter each way. It appeared, however, from the evidence before the committee, that Thompson received but little, if anything, from this source as the letters were deposited in the post office, and most persons to whom they were addressed demanded and received them without payment of $1 per letter, which the postmaster tried to secure for the carrier as a compensation for his services.

"By reference to a letter of the Postmaster General it appears that Thompson carried the mails two quarters in the year 1856, and two in the year 1867, although the latter quarter of the year 1857 was during the time the contract was awarded to J. B. Crandell. Joe (Crandell) was fully paid by the United States Government for that quarter, although it seems he failed to pay Thompson, who performed the service for him.

"It further appears, by the postmaster's returns in the auditor's office, that he paid Thompson postages collected at the office, in the sum of $80.22; and so far as the evidence goes, this seems to be the full amount of money he has received for his services.

"The committee, therefore, recommends that J. A. Thompson be paid for three-fourths of a year service in carrying the mails at the rate of $1,000 per annum, making the sum of $750, from which is to be deducted $80.22 already paid, leaving the sum of $669.78 still due," the report concluded.

This monetary settlement was approved by the Senate and sent to the House, which on Feb. 4, 1873, referred it to the Committee on Claims. There it was to die—but not before Snow-shoe had waded through mountainous snows to Washington in an unsuccessful attempt to breath new life into his compensation claim, as reported in the E. Clampus Vitus biography, as follows:

In mid-January, 1874, Thompson snow-shoed out of his Alpine County headquarters across the Nevada border to Carson Valley, where he took a stage to Carson City. There he boarded the cars for Reno, and, Jan. 17, was enroute to Washington via the Central Pacific. Three days later the train became stopped in a snowdrift 35 miles from Laramie, Wyoming, and there it stayed despite the efforts of four extra locomotives to move it.

Thompson became impatient and, in company with a fellow traveler named Rufus Turner, set out on foot. At Laramie the thermometer was ranging between 15 and 30 degrees below zero. Turner concluded he wanted no more pedestrian exercise. But Thompson was in his element. He pushed on alone and walked down the track 56 miles to Cheyenne, boarding an eastbound train for the Missouri River, where he was the first man directly through from the Pacific Coast in two weeks. Eastern newspapers acclaimed him as, "the first man to have ever beaten the iron horse on so long a stretch."

Despite the favorable advance publicity, Thompson's trip to Washington was unsuccessful. He never did receive the $669.78 the Senate had voted as still due him, let alone the $6,000 he was after.

Thompson returned to his Diamond Valley Ranch in Alpine County, where he died May 15, 1876. Even with the grim reaper approaching, his spirit remained undaunted. So weak he could not walk about to sow his fields, Thompson saddled a mount and broadcast his wheat from horseback.

An account of Snow-shoe was written by Dan de Quille, in which the noted Nevada journalist had the following to say:

"Thompson was the father of all the race of snow-shoers in the Sierra Nevada Mountains and in those mountains he was the pioneer of the pack train, the stage coach and the locomotive. On the Pacific Coast his equal in his own peculiar line will probably never again be seen. It would be hard to find another man combining his courage and sinews controlled by such a will. As an explorer in the arctic regions, he would have achieved world-wide fame. Less courage than he each winter displayed amid the mountains has secured for hundreds the hero's crown. To the ordinary man, there is something terrible in the wild winter storms that often sweep through the Sierra, but the louder the gale, the higher rose the courage of Thompson. He did not fear to beard the storm king in his own mountain fastness and strongholds. Within his breast lived and burned the spirit of old Vikings. It was this inherited spirit of his daring ancestors that impelled him to embark on difficult and dangerous enterprises.

This spirit incited him to defy the wildest rage of the elements."

Judge John S. Childs was Snow-shoe's partner in an overland stage line in 1858. Childs was Nevada's first divorce court judge.

After Snow-shoe Thompson died his feats lived on and grew into interesting legends of California and Nevada. Writers extolled his deeds and to later generations he became a veritable superman.

In Carthay Circle Theater in Los Angeles stands a huge granite boulder, taken from the hills near old Hangtown, as Placerville was called in early times, and a Sequoia tree transplanted from Humboldt County. It is a fitting memorial to the onetime Viking of Silver Mountain. On a plaque, one reads this inscription to John A. Thompson:

SNOWSHOE THOMPSON
A pioneer of the Sierras, who for 20 winters carried the
mail over the mountains to isolated camps, rescued the
lost and giving succor to those in need along the way.
Born 1826   Died 1876

The man thus honored, however, rests today in a quaint and seldom-visited cemetery near Genoa. A modest tombstone was erected by his widow 75 years ago, with a pair of snow-shoes carved above a truly prophetic benediction.

It reads:

In Memory of
JOHN A. THOMSON*

Native of Norway
Departed this life
May 15, 1876
Gone but not forgotten

Snow-shoe was honored at the theater with a memorial plaque and a "P" in the middle of his name. He probably should have been called the greatest of them all. He was to friends who witnessed his feats. Among these was W.R. Merrill, the postmaster at Woodford's, who wrote in later years, "He at one time went up back of Genoa on a mountain, on his snow-shoes and made a jump of 80 feet without a break."

John A. (Snow-shoe) Thompson may not have been the swiftest of the longboard riders, according to standards set by the Plumas and Sierra boys in the next chapter, but in Alpine County it was universally conceded that his daring exploitations on snow-shoes had changed the course of history and in many ways he was the greatest snow-shoer of them all.

*This spelling was a mistake, as historians have proved.

References:

First 80 Years of Skiing, by Bill Berry, Skiers Almanac, 1978

Territorial Enterprise, February 13, 1876, (negatives) "Snow-shoe Thompson," by Dan de Quille

History of Snow-shoe Thompson, State Sen. Swift Berry

Explorations Across the Great Basin of the Territory of Utah, 1859, Capt. J.H. Simpson

Crossing the Sierra, Hutching's California Magazine, February, 1857.

Eulogy, Snow-shoe Thompson, Territorial Enterprise, May 19, 1876

Snow-shoe's application for U.S. Citizenship

Carl Messelt, organizer of the Annual Snow-shoe Thompson Memorial Cross County Ski Race and 1946 president of the Douglas County Ski club, pays tribute to the memory of Snow-shoe Thompson.

Norwegian Consul General, Jorgen Galbe, places flowers on the grave of John A. "Snow-shoe" Thompson at Genoa, Nevada in 1946 during the first Snow-shoe Thompson Memorial Cross Country Ski Race. From left to right, Carl Messelt, President, Douglas County Ski club; Effie Mona Mack, President, Nevada Historical Society; a Nevada minister, Jorgen Galbe; Hans Wolfe, Reno banker; Fred Tuttle, Auburn; Richard Rowley, Reno; Bjorn Lie, Norway; Martin Espinal, Plumas County; Nowegian Einar Skinnerland of Telemark, Norway, race winner; Olaf Blodger, Auburn; Dale Gilbert, Reno and Andy Blodger, Auburn Ski Club.

Snow-shoe Thompson monument in Telemark, Norway. Left to right, Norwegian historian, Jakob Vaage and Bill Berry, USSA historian, 1974.

8

A Silver Link

"The La Porters were honored by the presence of Snow-shoe Thompson... at our races, which can only be beat by lightning. He runs behind like an ox team after a railroad car. He came out over 200 feet behind the last man in his squad."—Quicksilver Charley Hendel The Downieville Mountain Messenger.

This custom-made silver trophy, a replica of Norway's famed Holmenkollen cup, was presented to the National Ski Association of America at San Francisco in 1950 by Jorgen Galbe, Royal Norwegian Consul General. It is engraved: "In commemoration of Snow-shoe Thompson", in hammered letters around the top.

Engraved winners are: Stig Bengter, 1950; John Gianotti, 1952; Lief Sommerseth, 1953 and Sheldon Varney, 1953. Varney retired the trophy in 1956, donating the cup to the Western America SkiSport Museum, Boreal Ridge. The cup is the museum's keystone, symbolizing California's ski history and its Norwegian roots.

lost sierra roasting!

Snow-shoe Thompson's ignominious defeat in the Alturas Snowshoe Club's 1869 tournament triggered a newspaper blizzard unequalled by any Sierra Nevada gale that ever howled across the Sacramento Valley into the Lost Sierra.

Blast after blast of printed tirades blew out of editorial sanctorums throughout the mountains. Opinion pieces taking one side or the other of the Snow-shoe Thompson challenge, which in reality was also a challenge to the Norwegians, boiled up in newspapers far and wide, erupting in such far-flung sheets as the Virginia City Territorial Enterprise, The Downieville Mountain Messenger, The La Porte Union, Alpine Chronicle and The Sacramento Union. Smouldering regional jealousies flared into ink-fed fury.

February, 1869, Snow-shoe Thompson, a lone Norwegian looking for action, departed from his Silver Mountain City home and set out on snow-shoes for the Alturas Snow-shoe Club challenge races held at La Porte. After reading the newspaper ads and the La Porte Union's front page featured verse, "Mountains In Winter," he had decided to step into his longboards and challenge the Plumas and Sierra County boys. His departure was optimistically heralded by the editor of his hometown paper, the Alpine County Chronicle.

Thompson was the editor's favored "man." so he was given a "puff" story in the Chronicle hailing his prowess as he departed.

Thompson blew into La Porte full of confidence and, sorrowfully, he was soon wafted out by Quicksilver

Charley Hendel, a top racer, who in his spare time, was the ski reporter for the Downieville Mountain Messenger. Hendel had the following to say in his ski column:

"The La Porters were honored by the presence of Snow-shoe Thompson, the famous and celebrated express carrier, who in early days carried the express and mail from Placerville to Carson City as our friends over there will testify. He is a stout athletic man, a Norwegian by birth, handles his shoes in good style, and, we believe, is hard to beat on a travel, in the long run—with an express bag of about 60 to 120 pounds weight. But at our races, which can only be beat by lightning, he runs behind like an ox team after a railroad car. He came out over 200 feet behind the last man in his squad. He says he has been running since his childhood, but such speed he never witnessed before. He came all the way from Alpine County, over 200 miles travel, part on his shoes, part in stages, part by railroad, part in a skiff across Lake Tahoe, to see the races, and, if possible, to carry away all the money. He went home with his brains and pockets full of dope recipes which our boys kindly furnished him."

To Quicksilver's report, The Messenger editor added his own comment:

Alturas Snow-shoe Club member and FWSA president, B.M. (Milt) Zimmerman, Reno, accepted the Snow-shoe Thompson Memorial Trophy from Norway. Photo taken in 1950 at a Reno Ski Club race.

AUTHOR'S NOTE:
The resulting havoc in the Snow-shoe Challenge controversy festered bad blood for generations, until an international act of goodwill, staged 81 years later, brought it to an end. I clearly recall that occasion.

The ceremony held sway, Jan. 6, 1950 as I stood and watched.

From Norway, a long arm of friendship had reached out across the sea, to extend a friendly hand which was warmly grasped by a snow-shoer from California's Lost Sierra.

Firmly clasped at the San Francisco rites, the two hands quickly closed the old breach. Permanent healing came with presentation of a silver link...in the form of a custom-made silver trophy.

Acting on behalf of his homeland's ski federation, Jorgen Galbe, Royal Norwegian Consul General, presented the classic Snow-shoe Thompson Memorial Trophy to the Far West Ski Association acting for the U.S. Ski Association of America. The cup is now on display at the Western American SkiSport Museum at Boreal Ridge, California.

It seemed fitting that the man present to accept the trophy on behalf of California and Nevada skiers was B.M. (Milt) Zimmerman, Reno. Not only was Zimmerman the FWSA president, but also a mountaineer and a member of the legendary Alturas Snow-shoe Club, La Porte, who had learned his skiing on longboard snow-shoes in the Lost Sierra.

"This silver cup, donated by the Norwegian Ski Federation, is a link between my mother country and yours," Galbe said.

"It commemorates John A. Thompson, a Norway native, who starting in 1856, forged a link between the civilization east of the Sierra and the expanding world west of the Sierra.

"We Norwegians feel this type of sport is the finest and best in the world. More than any other sport it develops the human body, not only physically, but mentally and spiritually."

Galbe said the Norwegian Ski Society, in cooperation with the FWSA, would sponsor a special observance of the 1956 centenary of Thompson's first Sierra crossing. Galbe then announced Norway's recognition of the planned 1956 Snow-shoe Thompson race as a world championship event. Galbe said Norway would send top competitors to participate. The race had been launched in 1946 year by myself and Carl Messelt of the Douglas County Ski Club. It was funded by Raymond I. Smith of Reno.

"SNOW- SHOE THOMPSON—We have often heard of the famous snow-shoe runner by the name of Thompson, but never knew until recently that he was a tangible—a living reality. Thompson had heard of our snow-shoe races, and to satisfy his curiosity he attended those recently held at La Porte. He "veni vidied" but did not "vici" a bit. In fact, when he saw some of our runners make a trip, he said he did not want any of this. There is no doubt but that Thompson is a good traveler on the snow, but he had the frankness to acknowledge when he saw the boys run, that he knew nothing about racing. The shoes brought along by Thompson were a curiosity. It is reported that they were turned up at both ends, about seven feet long, convex on the bottom, and innocent of any acquaintance with "dope." In fact he never knew what "dope" was until he saw it at the races. The Sacramento Union gave notice that Mr. T. was coming up to put our boys on their mettle. Will it now state that Thompson was an extensive failure as a racer?"

snow-shoe jabs back

Thompson and a copy of The Messenger arrived almost simultaneously at Silver Mountain, where he and the Alpine Chronicle editor soon put their heads together.

TO: SACRAMENTO DAILY UNION

* "THOMPSON CHALLENGES THE SIERRA BOYS—The following challenge is offered on behalf of Snow-shoe Thompson and the Alpine boys, to the young men of Sierra, who must take up the gauntlet or forever hold their peace. Thompson is not scared at the "dope" much, and means business. If the Sierra boys like to have a chance at a big mountain, Thompson will accommodate them. We quote from the Alpine Chronicle of March 20:

"The Downieville Messenger says I heard of snow-shoe races, and to satisfy my curiosity I attended those in La Porte.

"I did go to La Porte, expecting to see some scientific snow-shoe racing, but I was disappointed. It was nothing but "dope racing," and is unworthy of the name of snow-shoeing. It is nothing more than a little improvement on coasting down hill on a handsled. The improvement is, that instead of uprights and crossbars from one runner to the other, they make their crotch answer this purpose, and they have no more control over their shoes than a boy has over his sled. They have exhibited some skill in making dope, but all they gain in this is that they make about the same time on a hill of 15 degrees that a man would, without dope on a hill of 30 degrees. These "dope riders" at La Porte are good, clever fellows, but thay have no more right to call themselves scientific snow-shoers than a man with steel skates on smooth ice, who with a spiked pole placed between his legs, thus pushing himself straight ahead, should be called a scientific skater.

"Now I, on behalf of the Alpine Boys, make these propositions to the Plumas and Sierra Boys, or "any other man" in the State:

"Come to Alpine County next winter and run with us. We will run you for $1,000 a side for each one of the following, viz:

"First—Against time; you select your hillside, and then we will select ours.

"Second—Side by side; we to select the hillside.

"Third—Over a precipice fifteen feet high, without use of poles, the one jumping the furthest, without falling, to take the purse.

"Fourth—From the top to the bottom of the highest and heaviest-timbered mountain we can find.

"Fifth—To run from the top of Silver Mountain Peak to the town of Silver Mountain. The altitude of the peak is 11,000 feet- -4,000 feet above town, and distant four miles. Now, boys of Plumas and Sierra, come over here: we will treat you well, and if you win our money you are welcome to it. If you come, be sure and bring that Messenger man along with you, and I will bet him $100, that if he attempts to follow me on snow-shoes for one day, he will break his neck before night.

"For the information of those who have not seen my snow-shoes, I will give the dimensions: They are nine feet long, turned up in front and flat-bottomed; 4 inches wide in front; 3- 1/2 inches behind and 1-1/2 inches thick in the center."

Alpine Chronicle responds with a left hook

TO: THE MOUNTAIN MESSENGER

"As the Messenger rather disparages our man, we will cite what he has done, and he can do it again for an object. Thompson, with a heavy bag upon his back, has frequently run three miles in five minutes; has jumped precipices and landed ninety feet, right side up, from the starting point; has command of his shoes to such an extent that on the steepest and heaviest timbered mountains he glides among the obstructions like the skater on ice; at ever so great a speed he will touch or pass within an inch of any designated object; he has often carried mail from Genoa to Placerville, 80 miles, in fifteen hours running time; he runs standing, and in coming down the mountains he does not check himself with the pole, but turns around and runs uphill when he wants to stop.

"Now, we have stated what one of our Alpine boys has done and can do, we hope this challenge will be accepted. We accept the wager of the $100 against our Messenger friend's neck."

the print war heats up

TO: SACRAMENTO DAILY UNION
(LETTER FROM LA PORTE)

"I see by your paper of the 26th instant that "Snow-shoe Thompson" and the Alpine Chronicle have been having a good deal to say about the late races of the Alturas Snow-shoe Club. Inasmuch as I was, in the absence of John Conly, the acting president of the club at the last meeting of same, when the celebrated snow-shoeist made his appearance, I claim the privilege of saying a few words on behalf of the club in reply. First, I will tell Thompson and The Alpine Chronicle man what the Alturas Snow-shoe Club was not organized for. It was not organized to make business for doctors and undertakers by running races down steep and thickly-timbered mountains over high precipices, nor for the purpose of encouraging gambling, otherwise we might be induced to accept Thompson's bombastic challenge, which, if he would come to time with the money, would be sure to put several thousand dollars in our treasury. But it was organized to fill in the time during the long, tedious winters when everybody is idle, affording an innocent amusement and health-giving exercise, thereby keeping the muscles in tune for the labors of summer. This being our object, we have always selected our tracks and managed our races with a view to safety, and invited all lovers of that kind of sport to come and take a chance for the purses which we have annually put up in accordance with this general invitation, as was supposed. This man Thompson made his appearance at our last annual races, was kindly treated in the most hospitable manner by the members of the club, beside being afforded every facility for entering into the sports of the day, such as "dope" and snow-shoes. How well he repaid our kind treatment his bombastic, breakneck article, copied in your paper, will show. *** Now I propose to compare the science, skill and ingenuity of this man Thompson in managing his snow-shoes with that of the boys of the club, as exhibited on the first day's racing, and which I challenge him to deny.

"On the first day, in the first race, among the 36 names who had entered was that of J. A. Thompson, of Silver Mountain, who had come all the way from Alpine County just to let the Plumas and Sierra boys know what he could do, and how easy it would be for him to gobble up that little dab of $600. The track being 1,450 feet long and 100 feet wide down a comparatively smooth hill, not very steep, Thompson being in the fourth squad, the drum tapped and down they came, Thompson a long distance behind, and, would you believe it, this man, who has such perfect command of his shoes that he can 'glide down the steepest mountains among the thickest timber with perfect ease, and run within an inch of any given object,' actually found this track of 100 feet, in an open space, too narrow for him, and landed, outside the ropes, much to the annoyance of the crowd of spectators there assembled, supposing themselves entirely out of the way of the most clumsy rider.

"Why, Doc Brewster has a mule that has been practicing this winter on snow-shoes that can beat him on an even string, and we have Chinamen that can discount him." (signed) MUDSILL.

TO: THE MOUNTAIN MESSENGER

"The challenge from Snow-shoe Thompson, with a 'P,' was evolved from a paragraph published some time ago in The Messenger. We presume Thompson felt somewhat chagrined at not being able to hold his own with our racers, but that's none of our business. If Thompson was ignorant of snow-shoe racing, and did not become sensible of his ignorance until he got a long ways from home, it is not our fault. Had Thompson applied to us for information before he left his Alpine heights to brave the 'cold world's scorn,' he would have got it at the lowest cash price. He rashly shouldered his goosenecks and meandered over into this section, the worst inflated man outside of an inebriate asylum. All this was not our fault, was it? We did not send word that someone wanted to 'see him,' did we? No! We just took his coming as people do with measles or smallpox—a thing to be suffered and borne meekly, as it may be in silence, considering ourselves fortunate with the reflection that people are seldom called upon to bear such afflictions twice.

"There was a time when we knew nothing about 'dope' when we, too, thought a snow-shoe should be flat bottomed and turn up about two feet in front. We, too, got our ideas from that same old geography, Thompson. But you see the universal Yankee that's "into" us soon cured us of all such foolish notions, and we cut off the goose-necks and dug a groove in the bottom just as you'll do, Mr. Thompson with a 'P' when you get a little more used to snow-shoeing."

TO: THE ALPINE CHRONICLE
April, 1869

"We are authorized to say that if Plumas or Sierra Boys will accept Thompson's challenge, the Alpine Boys will run them in either of their own counties, with or without dope, provided they are allowed to select the mountains, and further another $1,000 a side to run on one shoe, with both feet on it. When the time arrives for next winter's sports these challenges must either be accepted or rejected for our friends will be on hand with the $6,000 and amply backed with outside bets."

TO: THE LA PORTE UNION

"That's perfectly satisfactory. Your boys will have an opportunity to come to Plumas."

TO: THE MOUNTAIN MESSENGER

"Stick to your goosenecks, Thompson, and don't, we beg, don't use dope, lest you learn in your old age that your flat bottoms are not what you have so long and so fondly deemed them, and that you must give way to grooved bottoms, dope and last but not least, Young America's Thoroughbred Snow-shoers."

THE LA PORTE UNION

"We do not blame Thompson and the Chronicle editor for taking exception to certain statements in The Messenger in reference to his shoes, etc., but if the great snow-shoeist had made his challenge while he was in La Porte, he would doubtless have found his man, in fact we are inclined to the belief that he knew such would be the case, and purposely waited until he was so far away that he supposed no one would 'see' him before he published it. Thompson is a little above the average on snow- shoes, but if he really thinks he can lay over our boys, either with or without dope, on hillside, plain, precipice or race track, he is the worst fooled man in the State, and without intentionally adopting the 'your another' style of argument, we would suggest to him that he be present at the next annual meeting of the Alturas Snow-shoe Club; and avail himself of the opportunity to prove to the public that his challenge as published was not intended for buncombe.

"If we are not greatly mistaken, Thompson has lengthened his shoes about a foot and a half since he was here, and there is no telling what other improvements he has adopted for he carefully examined a number of shoes while stopping in this place. Thompson, how's your dope?

"The Editor of the Chronicle is in the same fix Thompson was—he don't know anything about dope, that's what's the matter with him. In chronicling the feats of Thompson he doubtless thinks he has published something passing strange: he probably imagines that Thompson is decidedly nifty on his little, old, flat-bottomed, nine-foot snow-shoes.

"Pshaw, man! Thompson's great feats as related, are nothing but boy's play when compared to what some of our snow-shoe sharps have and can do. We reckon you never saw any done, never saw how it works; you haven't found out that 'dope is king.'

"Why, Mr. Chronicle man, our boys with a hundred-pound pack on their back, often run six miles in five minutes and half the distance uphill. Jump precipices? Some of our ten-year-old boys call it nice sport to run on their shoes over the edge of Table Rock, turn a dozen somersaults (for be it know, it is over 300 feet perpendicular from the jumping place to where they land) and any boy who missed landing right side up with care, would be laughed at and called 'Snow-shoe Thompson's Boy.'

"Jumping chasms and ravines is also fine pastime for the youngsters, gives them a good start and lets them get under moderate headway, and they can easily jump across a ravine a hundred and fifty-two-and-a-half feet wide, and this on a dead level. What d'ye think o' that? Talk about command of shoes! Why that's "our boys" strong suit, that's the "jigger" that wins. Thompson ought to know that, if you don't. He has seen them tried.

"Your man, Mr. Chronicle, if he can stop his shoes by running uphill, has a little the advantage over us. On that point he lays over. Our boys use dope, they do, and oftimes when running down a steep hill, they jump with their shoes, right about face when in the air, and when they land, the shoes (of course they're doped) commence running uphill and unless the rider uses his pole, he can't stop them. It is only a short time since that one poor fellow lost his pole while making the jump and as he couldn't stop the shoes, he had to jump off; the shoes kept on and the last he saw or heard of them they were still "on the go."

"Those shoes were doped with common "traveling dope." But enough, you are not initiated in the mysteries of dope and if we told you but a half of what our boys can do on snow-shoes you might think we were "yarning" a little. Just remember hereafter, that "our boys" use DOPE."

TO: SACRAMENTO DAILY UNION
"Snow-shoe Thompson, AGAIN!—We have received a long communication from J.A. Thompson in reply to "Mudsill."

"As the Sierra and Plumas boys do not like the terms of his former challenge, Thompson comes at them again and says:

"Now, Mr. "Mudsill," I am about to make you another challenge, and see if you mean business or not. This will be to see which is the most science on snow, and which has the best command of his shoes, and I will confine myself to your kind of track; it shall be 1,450 feet long, 100 feet wide and a grade of 16 degrees. I will bet $100 on each run that I can perform them, or lose my money.

"First Run—We will make the track 25 feet wide, by stretching ropes on both sides, and I will go down through blindfolded; and if I touch the ropes before I get through, or fall down, I lose the money.

"Second Run—The ropes to be closed, leaving a space only three feet apart. I will run down between and not touch, and have one eye open.

"Third Run—Stakes to be stuck in a straight line every one hundred feet throught the track. I will leave the first one to the right, the next one to the left, and so on until I get through.

"Fourth Run—A stake to be within twenty-five feet from the outcome and turn around the stake; and when my shoes stop running my back shall be turned towards the outcome and my face uphill.

"Fifth Run—Two small stakes to be stuck at the outcome three feet apart, and I will run over them, one by each shoe, first going over one jump five feet high.

"Sixth Run—Jumps to be made every two hundred feet by piling snow up; each jump to be five feet high, and I will run over them all.

"Seventh Run—A log two feet in diameter to be drawn across the track at the outcome and sawed in two, and then drawn apart so as to leave a space but three feet between the ends of the logs, forming a kind of devil's gate, that if a man do not strike straight between them it will hurt him. I will go over one jump in this run.

"Eighth Run—The logs to be closed to within 18

inches of each other. I shall go through only with one shoe, the other to over the end of the logs.

"I will go through all these runs, and you can have the privilege of doping my shoes with lightning dope. Each run is to be timed; and if you think you can go through all these runs in the same time, I will bet you $100 on each run; but you need not run unless you want to. This is the only way I know of to make a challenge where you can see what science there is in snow-shoeing, without laying yourselves liable to fall in the hands of the undertaker. I have no doubt that you will accept this challenge; for you say I could not run through a track 100 feet wide, and here I agree to run on one three feet wide.

"I will not ask you to come to Alpine and run, and I do not want you to ask me to come to Plumas; but I will meet you half way, which is always considered fair among men.

"I name Cisco, on the Central Pacific Railroad, as a suitable place. There is plenty of snow and good hills, and being on the railroad, it will give everybody a chance to see the contest for the championship of snow-shoers.

(The 1939 U.S. National Downhill & Slalom Championships were held at Cisco Grove.)

"If the people of Cisco and the railroad company will get us up a belt, we will run for it, and the man that makes the best time takes the belt.

"If you men of Plumas and Sierra Counties, who call yourselves snow-shoers, will accept this challenge, we will put up $100 as forfeit in the hands of the proprietor of the Railroad Hotel in Cisco on the First of January next, the running to be on the 22nd of February."

TO: THE ALPINE CHRONICLE

"On Sunday last, for the first time, Thompson used dope and made a run of 1400 feet in 15 seconds: and on Monday last he went 300 feet higher and ran it in 22 seconds; he then again mounted higher and ran 2000 feet in 21 seconds. Those were not straight shoots, but serpentine and amidst rocks. He made the runs standing. The quickest ever made on the La Porte Track was 1480 feet in about 19 seconds."

TO: LA PORTE UNION

"Alas! and has it come to this? Is it possible Thompson, the man who was so "disgusted" with "dope riders" of Plumas, has condescended to ride dope and then brag of his performances with it? How has the mighty fallen! Thompson has become a "dope rider" Well, that is rich. He will have the opportunity to show how much improvement he has made since his appearance in La Porte in February last."

TO: LA PORTE UNION
Snow-shoe Challenge from the Table Rock Snow-shoe Club

"MR. SNOW-SHOE THOMPSON— Dear Sir: We find in the Sacramento Union your challenge to the boys of Plumas and Sierra counties.

"If you are serious in the matter and would like to put up a sufficient sum as forfeiture, we think we can fully accommodate you. We are somewhat doubtful of your perfect good faith in making these propositions by reason of certain circumstances going to show your greater desire for the expenditure of "wind" than collateral." * * *

"It is usual where a party challenges another to meet them on the ground of the challenging party, that they make provision for the extra expense incurred by the challenged party accepting. You have done nothing of this kind, but after considering that you have the best of it enough to challenge us, very coolly say: 'Come down to my home next year and I will run you even up,' and I will select all the tracks, and designate the manner of racing--I will, in short, run the "jigger" my way, and then if you win, we won't feel hard at you. ***

"We propose doing better than that by you. We wish to submit in the contest, the questions of speed, control of shoes, science, graceful and daring riding. We do not puff ourselves by saying we can beat everybody, but will say that we will contest with anybody, and let others decide, always excepting the Plumas and Sierra boys.

"Now we will say in all good nature and kindness, that we do not wish to detract from your world-wide reputation as a snow-shoeist or traveler; nor, are we vain of our own acquirements in that direction; still, inasmuch as you propose trying your skill against all the wide and wicked world, and Plumas and Sierra in particular, we propose to assume that you are in earnest, and to afford you the opportunity of going a "coasting downhill" all the way from a quarter of a mile, up to four miles, when, of course, we "have no more control over our shoes than a boy over a handsled riding down hill," you with your great "jumpativeness" and remarkable expertness, must come off an easy victor.

"Some persons might dissent from your opinion that we are not scientific riders and retort that straddling out ones legs three feet apart and leaning back at an angle of forty-five degrees on a nine foot pole dragging behind in the snow, as a steadying and steering measure, would scarcely deserve the title of scientific, but rather, that sitting squarely on one's shoes, with nothing only the least possible friction to impede the velocity, with no pole dragging behind to steady or guide the rider, would be full as graceful, and a far greater display of science. But these matters constitute only a difference of opinion between you and us. If the former style suits you we won't find any fault. Faultfinding, argumentative people may say that you are venting a little spleen when you kindly allude to us as only "dope riders," and will say they greatly wonder at the anxiety you manifest to secure some fast dope recipes if it would not do you proud to be able to go shooting through the poles as we "dope riders" sometimes do.

"Now, Mr. Thompson, we are going to make a few propositions, at the same time allow us to say that we are but a few of the many who are of the genus dope riders, and of whom we assume not to be the greatest. Undoubtedly there are many clubs in Plumas and Sierra who can and will offer you greater inducements than we possibly can. We propose to you this, that you come up here to Howland Flat and we will pay your expenses, coming and going, and we will run you for five thousand dollars on your five several propositions at one thousand dollars each, as hereinafter stated.

"We will use you becoming as a gentleman, and if you win our money we will publicly acknowledge our inferiority and immediately commence imitating your style of sliding; on the other hand, should it occur that we come out on top, we would expect to see you squat on your shoes and by practice become a dope rider. We say we would take pleasure in your doing this, we would put on your shoes our fastest "wax" well knowing, as we do, that you possess, at least, some natural talent in that line. You will please observe that in the following acceptance of your propositions we allow you to do all the selecting, and we will also observe that you have gone far away from anything like racing rules, both of which in any contest is very unusual.

"Your first proposition reads, "Against time, you to select your hillside, and then we will select ours," — We accept this, and will mutually agree as to distance.

"Your second, "Side by side, we to select the hillside." — We also agree to that, and will agree on distance as before stated and course of track.

"Your third, "Over a precipice fifteen feet high without aid of a pole, the one jumping the farthest without falling to take the purse." — If you mean by the foregoing to run down a hillside, and at some point on the track to go over a fifteen foot high jump, or precipice, while under motion, we also accept, both to go from the same starting point.

"Your fourth, "From the top to the bottom of the highest and heaviest timbered mountain we can find." — Though evidently intended to terrify, we will gladly accept, and will lay out the course of track jointly.

"Your fifth, "To run from the top of Silver Mountain Peak to the town of Silver Mountain, and distance of four miles," — To this we will say you can make the same choice here, and can find as lofty peaks to run from as you may desire, Pilot Peak of Mt. Fillmore, will, we think, afford ample grade and distance to suit.

"Now we have come up to your propositions, and will add a few, to each of which we will put up a forfeiture of two hundered and fifty dollars to run for a thousand.

"First — To run a track not less than two thousand

feet in length, side by side, grade of track to be not less than 20 degrees.

"Second — On same track as before, with shoes made alike and doped alike by us.

"Third — A race of two thousand feet on a grade of from one to fifteen degrees, side by side.

"Fourth — Over a track three thousand feet in length, side by side, we to give you twenty feet at the start, grade from ten to fifty degrees.

"Fifth — Up a grade of ten degrees one thousand feet long, with doped shoes, both pairs to be doped by us with the same dope.

"Sixth — Over a track of from ten to twenty degrees, one thousand feet long, without a pole, with ten jumps at the lower end, three feet high and thirty feet apart, with shoes made precisely alike and doped alike.

"Seventh — Down a track five hundred feet in length, of from two to twenty degrees, blindfolded, and without a pole, and to run between winning poles sixty feet apart.

"Eighth — On a track of from ten to forty degrees, two men on one pair of shoes with only one pair of toe straps, and only the smooth surface of the width of 4-1/2 inches shoes to stand on for one rider. No corks, nails, or other contrivance to be used to stick on with. Track to be not less than seven hundred feet.

"Ninth — Over a course two thousand feet in length, without dragging or touching a pole on the snow after one hundred feet from the starting poles. Speed and graceful riding combined to decide the race. The editor of the Alpine Chronicle to be a judge on your part, and the editor of the La Porte Union to be a judge on our part, and they to choose the third from editors who never witnessed a snow-shoe race.

"Tenth — Over a track, side by side, fifteen hundred feet in length, passing between two winning poles eighty feet apart, and each rider to describe a circle around the pole on the side of the track on which he starts. The one passing between the said poles first the second time, to be declared the winner. The riders not to carry poles.

"Eleventh — From two to four riders against a like number, as you designate, to run on any track that you may select of not more than half a mile in length where it is down grade. We to allow each of your riders one hundred dollars for traveling expenses.

"We will add that the amount here represented, together with the acceptance of our propositions, in all sixteen thousand dollars, constitutes but a fraction of the amount you can carry off if you can beat us.

"You will observe that we make eleven propositions and you five, and that there is money in them to the amount of sixteen thousand dollars, well worth anyone's trouble of going after, but more especially do they commend themselves to a snow-shoe sport. Not one of us are but working men making snow-shoes no part of our business and but that you insist on contesting with us you might have worn your laurels with never failing grace and dignity. But as it now is, it must be proved to us that they belong to you by right. We offer you these chances in connection with the acceptance of your challenge, in good faith, intending to win if we can by honorable means, and not otherwise. We will endeavor to make it an honorable contest, feeling that assured that you will do nothing but what looks to a like result.

"We would finally suggest that without delay there be sufficient sums put up by each party as forfeiture money to the end that each may know that the other is in earnest.

"With all due regards for you and the "Alpine Boys," and a pledge for your courteous treatment both in Plumas and Sierra, should you do us the honor of paying us a professional visit, and hoping soon to hear from you, we have the honor to be,

Yours truly,
TABLE ROCK SNOW-
SHOE CLUB"

Howland Flat   D. H. Osborne, Pres.
Sierra County, Calif.   Sam Wheeler, Secty.

TO: THE LA PORTE UNION

"The "dander" of the Chronicle man was raised by the comments of the Union on Thompson's first challenge which the Table Rock Club accepted, and he pitches into us because we poked fun at his scientific snow-shoer. The Chronicle also goes after

"Mudsill," but as that is not "our fight" we pass it by with the remark that our riders and their friends are as little troubled about "accidents," "Business for doctors," or the fear of "encouraging gambling" by giving purses to snow-shoeists, as Thompson or any of the Alpine Boys are."

THE ALPINE CHRONICLE
"Plumas can beat Alpine in the matter of "dope riding" and "jumping hogs," but we have "got 'em" in the matter of scientific "snow-shoe racing" and "racing hog." Silver Mountain can boast of having a hog that is "some on the trot," Every time her owner mounts his horse and starts out for a ride, she trots off after him, and no matter how fast the horse trots, she keeps close to him for a long distance. It is a comical sight—as much so as it must be to witness the evolution of the Plumas and Sierra "dope riders."

THE LA PORTE UNION
It is doubtless a comical sight, but it cannot equal Thompson's performance on the La Porte track. Why, bless you Mr. Chronicle, some of the spectators who witnessed it have to indulge in a hearty laugh every time the scene is recalled to their memory."

THE ALPINE CHRONICLE
"We will now drop this subject until next winter, when our Northern friends—if they are smart enough—will have an opportunity to walk off with an Alpine silver "brick," and in the meantime we trust their claims will pan out well, so as to enable them to accept Thompson's challenge."

February 24, 1946, after the first running of the Douglas County Ski club's John A. (Snow-shoe) Thompson Memorial Race at Spooner Summit, Nevada. From left, Andy Blodger, Auburn; Martin Espinal, Plumas County; Richard Rowley, Reno; Fred Tuttle, Auburn; Hans Wolf, Reno; Dale Gilbert, Auburn; Major Fraser West, Reno; Einar Skinnerland, Norway, race winner, and second place finisher Bjorn Lie, Norway.

THE LA PORTE UNION

"All right, Mr. Chronicle, you and the Alpine Boys must keep on thinking so; and if you come up here next winter to witness the contest you will be more disappointed and more surprised than Thompson was."

the wake of era newspapers

What happened in response to all this journalistic oneupmanship?

In Plumas County the wind blew and blew hard. Sparks flew straight down La Porte's main street kindling a blaze in the office of the Union. The fire raged unchecked and the newspaper plant was devastated. The La Porte Union appeared no more.

Over in Alpine County, the Chronicle was soon picked up bodily and moved all the way to Bodie, the then latest roaring camp of Nevada.

Each side had been fully heard. there was merit on both sides of the controversy, although it was never decided for gold in the silvery snow.

Seldom, however, did Thompson give such exhibitions for the mere name and fame of doing difficult and daring things. Many times, running down the slopes of 10,000-foot-high Silver Mountain, had Thompson made clear leaps of 50 to 60 feet, but his greatest feats and flights on snow-shoes were only for the eyes of his closest friends.

Snow-shoe Thompson, the pride of Norway, had been hit and Norwegian pride hurt by the editorial storm, creating a rift that was not healed until Jorgen Galbe, Norway's Royal Ambassador General in 1950 healed it with a Silver Link, the Holmenkollen Cup.

References::

The Author's notes.
The La Porte Union, La Porte, CA.
The Downieville Mountain Messenger, Downieville, CA
The Sacramento Union, Sacramento, CA
The Alpine Chronicle, Silver Mountain, Alpine County, CA

Snow-shoe Thompson Memorial Cross Country Race winner, Sheldon Varney, shoves off. The 1956 event ran exactly 100 years after the famed mail carrier's 90-mile trek from Placerville, California to Genoa, Nevada across the Sierra Nevada on snow-shoes with a 40-pound mail sack on his back.

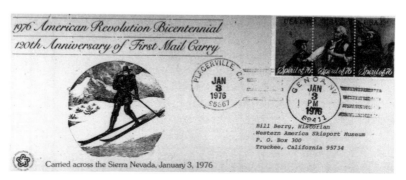

1976 Bicentennial celebrations in the High Sierra included paying respects to Snow-shoe Thompson, who represented as well as anyone could the American pioneering spirit. Middle photo, the sons of Norway gather at Snow-shoe Thompson's grave January 1976.

Bottom photo, first U.S. bicentennial observance was staged at Genoa, Nevada, January 3, 1976. Included, from left, John Watson, FWSA president; Vicki Nash, Nevada, ARB Commissioner and John Warner, Washington D.C., Administrator, American Revolution Bicentennial. Later elected Virginia Senator.

9

Dope is King

"They have attained a speed of over 200 feet per second. Upon reaching the bottom of the declivity the riders often report they have not had time to breath between the start and the finish."
— THE SNOW-SHOERS OF PLUMAS BY H. G. SQUIER, FEBRUARY 1894, CALIFORNIA ILLUSTRATED MAGAZINE.

the science of dope

Longboard kings were made and broken by the mysterious interaction of "dopes" with the temperature and texture of the snow they ran on in the days of the Snow-shoe Erawhen dope was king.

From the pages of the La Porte Union comes the story of "dopes," as racers arrived, Feb. 22, 1869, at La Porte, for the Third Annual Racing Meeting of the Alturas Snow-shoe Club:

"From each camp came from one to a dozen riders, packing on their shoulders a pair of polished, trim and favorite 'racers.' Some of the shoes were beauties; each town seemed to have a shoe different from the others, and as a matter of course, the peculiarities, advantages or disadvantages of the different makes was duly criticized and commented upon. Some were thin and limber as whale-bone, others thick; some had narrow grooves, others wide ones, etc.

"With the delegation from each town came a sack, and in that sack was the "Dope." The sacks varied in size; some were as large as good-sized grain bags, while others would not hold more than could be contained in a flour sack. Each delegation had its dope-man, whose business it was to furnish the riders with the best wax to suit

Dopes, the secret of lightning speed on snow-shoes are shown, above, in front of the Williams' family dope box. The Alturas Snow-shoe Club racing star, sewn on red silk, is draped over the collection of ingredients.

the snow, and to guard the dope so that no outsider could procure a sample. The care and watchfulness bestowed by a mother upon her first-born could not be greater than that of the man entrusted with the dope.

"As the snow is supposed to be different in each locality, it was necessary to find out what dope was best suited for the snow on the race track, consequently, during the day before the races, small squads of men from various localities were scattered all over the hill, testing the quality of their mixtures."

Race results in the Union were well-sprinkled with quotes establishing the fact that having used skis with the right dope on the running surface, separated winners from losers among top favorites in the "saw-off" heats and finals. Phrases from the La Porte Union included: "Howland Flat dope proved too fast for Spanish Flat... Charley Lyttick, Howland Flat, gave his wooden war-steeds an extra polish and passed the poles ahead of "Yank" Brown, winning the race and money...Port Wine and La Porte had not taken a trick, and full delegations were out to take advantage of the day to try dope, but, alas, it availed nothing...In the saw off, Gould, Stone and Squires, all of Gibsonville, but running on Saw Pit dope, won their squads...Howland Flat and Poker Flat dope cleaned the platter, won every race and captured the money."

Dope, the mysterious mixture of spermaceti, oil of spruce, balsam fir, pitch pine pitch, camphor and oil of tar propelled winners across the snow with lightning speed.

Isaac Frank Steward, known as Frank, of La Porte became a towering figure among the many snow-shoers; participating in the sport's important phases, including, racing, dopemaking and longboard manufac-

Frank Steward, at right, posed for this photo in 1911 after his champion, Joe Bustillos, the racer at left, won the purse, using another man's dope. From left, Bustillos, George Burelle, Ray Cayot, Lester Hillman and Steward. W.B.B. Collection.

turing. None seems more favorably known than the Down East Yankee as his exploits were reported down through the years of the Snow-shoe Era.

Steward's career encompassed the Snow-shoe Era. He arrived in Rabbit Creek in 1851 as an 18-year-old gold seeker out of Skowhegan, Maine. He died in La Porte in 1913, at age 62. He was in Rabbit Creek for the legendary start of snow-shoe racing in 1857. He saw his community's name changed from Rabbit Creek to La Porte. As a top snow-shoe racer he quickly joined the new Alturas Snow-shoe Club when it took ski racing into an organized sport in 1867.

Steward was one of the longboard riders who bested Snow-shoe Thompson when he came to La Porte, Feb. 22, 1869, for the Third Annual Alturas Snow-shoe Club Meet. Steward was among 36 entrants in that famous race, in which the final winner of the main race purse was La Porte's O. Liberty.

Steward outlived the 34 other racers who might have also claimed to have defeated the redoubtable Alpine County Norseman. Because he survived all the others, the legend grew around the La Porte racing circuit that it was Steward who actually outsped Thompson. Such legends die hard.

In his own particular domain, Steward was to become no less famed than his Norway-born rival. While Thompson died before the sport was developed in other parts of the world on an organized basis, Steward lived to see the formation of the National Ski Association of America, even though he was practically ignored by it, because of the world's blissful ignorance of the Lost Sierra.

Nevertheless, Steward became a dominant figure of the snow-shoe sport, first as La Porte champion, later as a manufacturer of snow-shoes and dopes. His prowess was quickly recognized in the very earliest snow-shoe days, before written accounts appeared. He enjoyed growing recognition as a member of the Alturas Snow-shoe Club. The club appeared in 1867 and bowed out with a final race in 1911.

Down through the years Steward's racing exploits earned him the title of "world champion," even though within the confines of the Lost Sierra.

Basking in the glory of his early-day feats and sporting a reputation for superb dopemaking, Steward became one of the region's most eligible and pursued bachelors. The chase ended in 1877 when he married Alice Pike, whose father, Charles Pike, first objected. Later the two men joined in the snow-shoe manufacturing business.

When Steward died in 1913, Alice married Thadius Holcomb and lived in Challenge, a hamlet just below the snow belt on the Sierra's Western slope.

In 1938 Alice told of her romance with the snow-shoe champion.

"I was 18 and Frank was 44,," she reminisced.

Alice Pike Steward Holcomb, pictured in 1938 holding a copy of Sports Illustrated which told Steward's story. She told of her romance with the legendary Frank Steward. After Steward died she wed Thadius Holcomb.

"We met at a snow-shoe dance. It was a romance of the ski trail, I guess."

Alice had been the belle of La Porte. At 79 she still retained the sparkle that had won Steward during the Snow-shoe Era.

"My father came from Yuba County in 1865," she recalled. "He was a carpenter and immediately began manufacturing snow-shoes. When I was married to Frank, he joined father in the business. Their snow-shoes were considered the best on the market. They cost $6 a pair and $1 for the single pole.

"But I should tell first about the times when I was a little girl. I was six when we moved to La Porte. I learned to snow-shoe that same year. I had to!

"The entire life of the community centered around snow-shoeing and preparations for the winter races. I can still visualize the riders coming down the hill. They came so fast you could hardly see them at the finish.

"Everyone had a good time at the races, but as a young girl I didn't much care for that excitement. It was the dances every night after the squads had been run that I liked the best. They would be held in the Union House. It was there that I met Frank. He was dark, handsome and had a beard that reached his chest."

Character-building associations were found by children in snow-shoeing, according to the little lady, who once had been married to the champion rider. Youngsters looked up to the older men who were dopemakers. They did not talk too much about the sport during summer months, but when autumn came and the dopemakers began gathering balsam and other ingredients to be mixed into secret preparations, they began youthful evaluation of the merits of the various older men.

"Frank wouldn't let anyone see him preparing his dopes," she recalled with a twinkle. "He worked with his balsam pots on the kitchen stove. I can still smell it brewing— balsam, camphor, paraffin, oil of hemlock— a good, clean smell. He made large cakes of it and had different recipes for every kind of snow and weather."

Mrs. Holcomb asserted that the dopemakers all came from good stock —American, English, Scottish, Welsh and Irish—and that few of them were drinking men, although many of the riders were known to touch a drop on occasion.

Time can be turned back in the Lost Sierra to tell about the dopemakers and their riders as they come across snows of many years ago.

In Saw Pit Flat and Onion Valley the art of snow-shoe making and dope brewing began in 1854, with "Old Buckskin" Porter being the wood expert, while Jack Clinch, the first recorded racer, was the first dope king. The team of Porter and Clinch sent such noted racers around the racing circuit as Robert Oliver, Robert and William Francis, Frenchy Riendeau and Bill Metcalf. In nearby Poorman's Creek was a snow-shoe man named Bill White, who picked up the combined art from Porter and Clinch. He later opened another shop in Gibsonville. It was in that way that snow-shoe manufacturing and dope-making business grew.

From the mid-fifties until 1885 in Gibsonville, which was the trading center for nearby Whiskey Diggings and Hepsidam, three dopemakers emerged into great prominence. They included Silas Jedkins, William Ross and later, Frank Saulter. In 1869 Jedkins and Ross turned over the dope end of the business to Saulter, who then became Gibsonville's undisputed dope king, while his partners continued to manufacture snow-shoes with White, who had started the business in the first place.

Saulter's riders were George Roubley, William Stone, Alex Ross, Tom Stevenson, Jacob Gould, Elias and Eugene Squier. All of them made racing history prior to 1883, when a new bunch of racers challenged their leadership. Among them was Matthew Judge, whose string of winning purses in the following seven years was never equalled. He won more money and races than any other rider in all the tournaments ever held.

Howland Flat's more noted dope men included Hiram Walker, a mining engineer and A. J. Howe, the town's leading lawyer. Their more noted riders were G. "Yank" Brown, winner, champion belt winner in 1869, and Tommy Todd, who set an all-time world speed record five years later. Walker and Howe moved away from Howland Flat and left their dope recipes to H.P. Stout and Patrick Kelleghan. The latter's proteges included such good Irish names as James Cosker and John Flournoy.

Dean of Poker Flat's combination dopemaker-racer was Charles A. Scott, Sr., who began his riding around 1860 and continued until 1883. March 18, 1883, Scott dropped out of a final squad race in order to properly prepare the boards for his younger protege, Tom Costello, who then went on to take the purse.

The first dopemaker in Port Wine was N.P. Abbott, better known as "Bonney," He was among the greatest riders of the very early days. His secret preparations were used by the other Port Winers, among whom were O. Liberty, winner of the Feb. 22, 1869, purse when Snowshoe Thompson was so badly beaten. Others using Abbott's dopes included T. Mason, J. Killey and Pell Tull. In later years of the Snow-shoeing Era, Lin Franz filled Abbott's shoes.

Dutchman's Ranch boasted Henry Ermantinger, who made both dope and snow-shoes throughout the late eighties until racing ended. He liked to ride the longboards and once introduced a pair of 14-foot snow propellors in La Porte. In 1893, he fell, going over a bench on the track in a saw-off with Louis Sibbly and John H. Barrett. That ended Ermantinger as a racer.

There are no records recalling the kings of early days in Scales and Poverty Hill, but R.H. Kingdon of that area made some fast dopes in 1910-11. His best rider was John Bean, a member of the 1950, revived Alturas Snow-shoe Club

The biggest surprise the dope men ever got was in March, 1892,when the Johnsville boys came to Gibsonville. Included in the riders who trekked across the mountains to challenge the old-timers' supremacy were John and Arthur Woodward and Reddy Gallagher.

John Woodward was a dope man, who had also made their shoes. La Porte entered 10 riders and Gibsonville, six. Taking every squad they ran in, the Johnsville boys cut out a lot of good riders. The final squad for the money was between Charley Hewitt, for John Madden of Gibsonville; James Mullen for Frank Steward of La Porte and the Johnsville boys. They got away to a good start, with Mullen the last away but the first over the finish line. Art Woodward hurtled across right at his heels. Only then did the high-mountain dope men breathe easier, and they never went out of their way to visit Johnsville again!

The Union House Hotel, La Porte, was referred to as the "first ski lodge in the West." It was at the Union House that Alice Pike met Frank Steward at a dance. Photo circa 1912.

La Porte not only had the first snow-shoe club, but the most dope men. They featured the most famous racing track as well. Its track was not the steepest, nor the longest, but in a run down it, Tommy Todd set the all-time record. Many lightning runs were featured on its track. By the turn of the century the great decline in mining and the subsequent drop in population reduced racing with specified purses to one La Porte tournament a year.

The town's racing dopes were tagged with such names as Harris, Williams, Mullen, Hillman—and Frank Steward.

Racing had been abandoned by Steward in 1880. Following that retirement he developed many riders to run on his mountain magic. These included Joe Pike, Ed Pike, Charley Yard, Ed Berry, Al Dubuque, Old Joe and Young Joe Bustillos, Mike Bustillos, Al Primeau, and Clyde, Cleve and Leonard O'Rourke. Mullen's dope was a family affair for Neal, John and Ed Mullen, while George, Louis and Bill Hillman were coached and snow-shoes doped by their old man John Hillman.

And here was a Williams family known for its "Old No. 25" snow magic.

In the final days of racing the entry lists were not what they had once been. Nevertheless the snow-laden winds of La Porte were still blowing roses onto the round cheeks of the mountain girls and courage and determination blew into the hearts of men when the championship of 1908 was decided.

Tensed at the starting line crouched Young Joe Bustillos, short, dark and thickset. Straining beside him leaned his taciturn brother, Old Joe. Both were running on Steward's dope. Between the brothers hung the determination to make that dope win again as it had often done before for the glory of Rabbit Creek.

Beside them stood big Jim Mullen, generally called "Colonel Bull," full-faced and square-jawed. His look said something must break before he will. He's running on the family dope.

Defending Champion Orville Squier, Gibsonville, stands anxiously on Ab Gould's dope. The youngest of the four, he is a poised mass of nerves and steel springs. The previous year Squier had wrested the championship from La Porte and taken it to Gibsonville, where he said, "It belonged."

The starter raises his hand—the racers' muscles tighten. Crack goes the pistol. In an instant, with furious strokes of the single poles, the four men drive themselves forward. In a moment they top the edge of the chute and plunge down the incline. Down, down they drop, gathering speed with every breath. Now they're flying, now falling; yes, literally falling through space. Guidance is blind instinct now, for they are going too fast to see; even the onlookers cannot distinguish who is in the lead.

Then, just as they drop below the incline and dart out toward the finish line there is a crack. It's clear as the starter's pistol shot. A snow-shoe has broken. One racer, none can tell who, goes tumbling, head over heals.

With the slowing speed as they flash through the finish poles, it can be seen that young Orville Squier is in the lead, winner by a scant foot over Shorty Bustillos and Mullen. Seconds later Old Joe picks himself out of a snow drift and waves to show he is not injured.

Pandemonium breaks out. Cheering echoes for Gibsonville and the champion. It's then time for one drink at the trackside bar, then on to dinner and the dance.

John G. Barrett, a well-to-do mining man and veteran of many race tournaments, is called to give a little talk.

There would have been no snow-shoes without a tip bender. The ski tip was pushed into the curved mold at bottom, held by a slat above, while hot steam, a hot iron or blowtorch was applied to set the tip's curve.—Plumas County Museum Photo.

Here is what he had to say:

"A great deal has been said about snow-shoe racing Some claim the old, original riders made faster time than the younger riders do now, but I can't quite agree with them. I think racing was at its peak from 1880 up to now. Boys raised and brought up on snow-shoes were better riders than the older ones. The dope and snow-shoes were vastly improved. It has been claimed that Tommy Todd's run of 1804 feet in 14 seconds was the fastest ever made. Well, maybe it was, but I remember very well a race in March of 1893 off the hill here. It was nearly dark and the stars were out when Bill Gould stepped up to Lon Sibbly and myself and asked both of us if we had any idea of the time made. He said it was 1886 feet in 13 seconds. The next winter here I came as a spectator and saw Andy Hewitt and John Moriarity come off the same track in 12 seconds. I think that was the fastest time ever made. Hewitt was riding Madden's dope and Moriarity for Lin Franz of Port Wine."

old no. 25

Members of the famed Williams family of snow-shoe racers from Johnsville, apply their "Old No. 25" snow magic dope before a 1911 race at La Porte. W.B.B. Collection.

We now leap through the years to March 11-12, 1911 to recall the all-time finale of snow-shoe racing staged by the Alturas Snow-shoe Club at La Porte.

Gathered to compete are 29 riders, all determined to win the four posted stakes.

Harold Pike knows of $67.50 in the opening event. Then Young Joe Bustillos cops $48, but breaks the heart of Frank Steward, the old master, in doing it. Times had become hard and money was then scarce in old La Porte. Young Joe wanted that cash badly, so much so that for the final ride he switched from Steward's "Old Black" magic dope for late afternoon running to the Williams family's "Old No. 25."

Steward's feelings were hurt badly by the dope shift. Bitter feelings developed that night among the riders and their sponsors. Old man Steward took to the bottle, something he had never been known to do before. In his cups he told the boys that, following "tomorrow's races," he would never dope again.

Ed Williams won the next day's opening purse. Then Steward had his final moment of triumph when Jim Mullen slid to victory. The track that day was 1500 feet and the Mullen-Steward combination had it in 14 seconds. Nevertheless, even in with his final triumph, Steward was as good as his word. He racked up his dozens of pairs of racing shoes for keeps and handed his preciously secret dope recipes over to Leonard O'Rourke who shared the two dope recipes below:

1. Frank Steward's Old Black Dope:

(First used at La Porte in 1869 and again in 1911. The cooking time remains a secret.)

2 oz spermaceti   1 tablespoon balsam fir
1/4 oz pine pitch   1 tablespoon oil of spruce
1/8 oz camphor

2. Johnie Williams' No. 21—a cold snow dope:

1 oz. spermaceti   1/2 teaspoon balsam
1/8 oz camphor   21 drops of oil of tar

Cook for 45 minutes, then add 1/2 teaspoon oil of tar & cool.

(This February 1, 1895 Williams dope recipe; "won five purses on the cold snow.")

Frank Steward

Two years later, Steward passed way. He was gently placed beneath his beloved snow for keeps.

Dopemaking was a science that affected the entire population, man, woman and child of the Lost Sierra during the entire Snow-shoe Era. There were as many kinds of dope as there are different kinds of Christian worshippers, perhaps more. Everybody ran on dope; they had dope with their meals, dope with their drinks and some got pretty well doped, too. There was dope in the mine tunnel and dope at the theater and if anything in this book makes the frequent use of this word as ridiculous to the readers as it has become to the author, it is hoped that the public's reading tastes will be improved thereby.

March 4, 1869, the dope question worked up Quicksilver Charley Hendel to a point where he wrote the editor of the Mountain Messenger as follows.

EDITOR, THE MESSENGER:
"Once again I am trying my almost rusty pen to break the long silence of your many correspondents of the northern part of old Sierra County, to give you a short synopsis of what is going on over here. By so doing all your former spicy correspondents may awake from their lethargy and tax their brains a little, to let their friends know that they are living, and running, on fast dope. By the way, I am happy to inform you that, for instance, "Cato the Younger," who drifted years ago toward the North Pole, and who is now dealing in most everything but cards, ran last fall, on fast dope, a splendid race (following in the same track as your fair lady correspondent 'Anita', who was the champion of lady snow-shoeists a few years ago), coming out winner at the `poles' of matrimony, carrying off as his prize from the Hill called Poverty, something attached

Johnsville, circa 1881, during the boom times at the Plumas/Eureka mine. The home of the champion snow-shoeing Woodward brothers can be seen at the extreme right.

to an apron-string. In consequence, I think he is excusable for being silent during this time (which married people call 'honeymoon'.) Well, my friend, Cato, I hope and wish you much success in gathering around you young Catos, who, if they want to be president, they may be at least like you, good runners on fast wax.

"Your fair lady correspondent 'Anita', it appears, has, since she arrived in the haven of matrimonial life, attended to more serious matters in her kitchen, while her bosom friend passes his time in shooting the icicles off somebody's flumes, much to the amusement of himself and of his friends.

"I have been informed that Uncle Ben Wallace had a pair of splendid racers made by the celebrated manufacturer Thomas Gibb and in order to try and win the races here, March 10-12, he is trying all kinds of dope fast enough to break somebody's shoes, for he tried it on his with good success. I must not forget in going through town to mention the St. Louis Wolf, for everybody knows him, and he knows how to manufacture the fastest of dope for thirsty souls, that is bound to win if a fellow gets enough of his compound. I must say right here, Mr. Editor, that I see you made a slight mistake a few weeks since in regard to that Wolf, for you credited his little offspring to one Wolf in Port Wine. I hope you will correct it, for it might cause some trouble in the family, for John says he will pull the wool from your eyes if you give him no credit for his work, although he lost one keg of lager on it because it turned out to be a girl instead of a boy.

"In going down around the corner from Hoop into Main Street, you pass the man of fat and pork, Carmichael, whose dope is fine for strangers after traveling a good distance. He says he is going to be on the hill, and show all creation what he can't do. A little further down you may find Ed Danforth, the dealer, (God knows at what?), who practices again with fast wax. After passing Charles Street, you stumble into a hotel called St. Louis Hotel, second to none in the state, except the Cosmopolitan, for be it known, it is kept by David Corbett, who knows how to smile with the hungry. Dave expects a big crowd at our next bamboo snowshoe race and has cut into requisition all the blankets and beds for his many guests. After passing Williams, in his dry water office, you will find Schwarts' Restaurant, the place where our boys will give all outsiders a chance to dance with our pretty girls. I will stop talking now about our townspeople or else I may say something in a turn where I would speak of our visiting lady from that Hill of Poverty, it be a delicate and dangerous field to enter, I will better leave town at once.

"Our Howland Flat neighbors will have next Monday week something, great, under the auspices of the Table Rock Snow-shoe Club, assisted by the Honorable Order of E Clampus Vitus, by permission granted by the G.R.G.I.A. of Mokelumne Hill, who will be present himself to see the fun—when, and where, it is supposed

Woodward family portrait taken circa 1898, shows famed snow-shoe brothers from Johnsville, from left, Edward Francis (Frank), Arthur and John. The Woodward sisters from left, includes Alice, who became veteran snow-shoe competitor Alice Meffley; Birdie, who became better known as Fannie Woodward Hunsinger and Florence. Photo courtesy of Birdie's granddaughter, Gwen Ramsey Scott.

everything will pass off satisfactorily. I hope the Downieville Clampers will turn out in a body, for such a running as they never saw. Their wants will no doubt be cared for. Send your candidates over and they will be elected to some office—if they are in good standing in the community. If they come over and can't ride fast dope, we will allow them to ride on their poles.

"The races at Port Wine passed off in splendid style, a full description you have no doubt, ere this, notice in the La Porte Union. The most noticed of all men in these races was Klekner the Great. He rode gallantly his pole all the way down, slow, but sure, giving all others a chance to win. Our friend Benjamin Barnes of the Alturas Club, who is now Federal Tax Collector, was quite excited, and busy, as one of the judges at the races. He said he was annoying his spouse at night time, dreaming of the merits of the different racers and all kinds of dope, imagining himself to be still on the track. Such, Mr. Ed., is the consequence of the excitement of snow-shoeing."

QUICKSILVER

St. Louis, March 4, 1869

References:

The La Porte Union
The Downieville Mountain Messenger
The 1911 Feather River Bulletin
San Francisco, 1937 interviews by the author with Mrs. Birdie Haun Swingle, Mrs. Alice Pike Steward Holcomb.
Susanville, late 1937 interview with Art Woodward.
Letter from John Barrett

Alturas Snow-shoe Club racers lined up at La Porte, March 12, 1910, following a weekend of racing. From left, Medev Bustillos, Al Primeau, Clyde O'Rourke, Harold Pike and Will Primeau. At right, dopeman C.E. Pike, Harold's Uncle. 1910 postcard photo.

10

Volunteer Corps

Winter tragedies in the mountains were a popular subject in the Sacramento press. Deaths were always front page news.

Death, destruction, broken bones and torn ligaments were commonplace during the Snow-shoe Era. Avalanches roared down from the heights to obliterate life and mining camps below. Freezing cold often weakened mail carriers, letting death slip up on them. The frequently broken lower leg, the tibia, became known as the snow-shoer's bone. Pioneer doctors were expert in its reconstruction. Dislocated shoulders, hip and knee joints were snapped back into place by volunteer experts, while the less seriously injured took care of their own ankle sprains and bruises.

Though many of the snow-shoers lived rough and talked rough, it was not true that an every-man-for-himself policy was universal throughout the mines.

Snow piled deep in the Lost Sierra, where before snow plows were invented, doctors had to travel on snow-shoes. In the above photo taken at La Porte in 1913, a homeowner, at right, tried to clear his shed roof. W.B.B. Collection.

Men were capable of feeling the sufferings of others and were moved to answer calls for help. This was particularly true of the pioneer doctors and county authorities, many of whom experienced suffering themselves in encounters with the snow.

Among the early physicians were Alemby Jump of Howland Flat, Edmund G. Bryant of Downieville, E. X. Willard of Port Wine and H. L. Willard of La Porte. In 1855 the Sierra County supervisors appointed Dr. Jump resident physician in his district, while two years later Dr. Bryant was named supervisor of the county hospital in Downieville. The Willard brothers were officials of the Port Wine and Alturas snow-shoe clubs and became the bone-setting specialists in their respective communities.

There were many other pioneer physicians, but the snow-shoers remember best the names of those four men. Jump and the Willard brothers traveled many miles through the snow to make their calls. They often set the bones of injured racers. The more seriously injured racers were then removed on sleds by volunteer corps to Downieville. There, Bryant would take over their cases at the county hospital.

It was said of Dr. Bryant, who was hospital supervisor from 1858 through 1862, that he never turned a deaf ear to the cry for help from any disabled or broken-down snow-shoer. This was remembered four years later when he died under tragic circumstances in La Porte. Residents passed the hat to enable his widow to pay pressing debts about town.

To enumerate even a small percentage of the deaths, accidents and destruction, which occurred during the Snow-shoe Era, would require a super-sized volume. Such incidents were so commonplace that a report in the news columns of the Plumas National in 1876 brushed aside the death of a snow-shoeing miner with the terse comment, "cases of this kind are not unusual with us in the mountains."

The winter of 1865 is a good one in which to begin picking up stories of accident victims. In mid-February the La Porte region went through the rigors of a four-day storm. A terrific gale shook the houses and blew snow into huge drifts. Snow-shoe travelers encountered great difficulties when the skies finally cleared.

From Port Wine the rounds of Dr. E. X. Willard immediately were resumed. While riding on his snowshoes, from the residence of a Queen City patient to the home of another at Eagle Tunnel, he tumbled in the deep snow, breaking a leg and fracturing an ankle. Word of the accident was relayed at once by a snowshoe messenger to the doctor's brother, H. L. Willard, in La Porte, who soon crossed several miles of rugged terrain to render medical attention to his brother.

The incident was reported in the Mountain Messenger and ended with a report that may be considered interesting reading for those who care for present day injured skiers.

"We are indebted," the reporter wrote, "To Dr. H. L. Willard for the following concise description of his brother's injuries: The tibia, or main bone in the leg, between the ankle and the knee, received a transverse fracture nearly mid-way; also, he dislocated the ankle joint of the right foot,

Snow was a constant challenge to doctors hurrying to emergencies as shown by this snow-covered bridge near Johnsville en route to town. Photo, circa 1910. W.B.B. Collection.

and entirely broke up the union of the tibia and bones of the foot, which allowed a suffusion of the sonovia among the lacerated integuments and parts contiguous to the injury. I consider the ankle injuries most liable to be an abiding obstacle in the way of his recovery."

Dr. Willard's report on a broken tibia may be considered to have been the father of all such others that are commonplace in these modern times among records kept at our nation's skiing centers.

Meanwhile the deep snow had created problems in other sections of the Lost Sierra.

Out of Howland Flat, the snow-shoes of Dr. Jump were pointed toward Five Points, where Alex Cowan had broken an arm when a huge drift sucked him in, while speeding downhill. It was thought, "the arm was broken by being brought into sudden contact with the balance pole."

Simultaneously at Morristown a Volunteer Corps was preparing to remove a man named Scribner to the hospital in Downieville. Scribner's collar bone had been broken when he crashed into a tree while practicing on the race track. As part of the report, The Downieville Mountain Messenger editor importuned the town's residents to "forward the details of the impending snow-shoe races at the earliest opportunity."

Apparently there were no extra snow-shoe messengers to spare as the requested report cannot be found in the old newspaper files.

All extra hands were organized to rescue four men reported lost in the vicinity of Mount Pleasant, where they had become snowbound while on a trip to inspect a new-fangled water wheel installed to operate mining machinery. Much anxiety was expressed that the men had perished in the snow, but the "lost lambs" were finally discovered roosting high in the wheelhouse. They were still safely gazing at the ponderous object of their journey.

Roaring avalanches killed many during the winter of 1868, a time when the snapping of bones was commonplace. In March, newspapers were filled with accounts of deaths and racing results. A huge snow slide on the Yuba's North Fork cascaded down upon the Independence Mine and Eureka Mine, killing two men. Three slides at Sierra Buttes wiped out small communities below, scattering a 24-stamp mill about the valley floor and tossing three miners to Kingdom Come. Among a rescue party organized in Downieville was John Corbett, county assessor. He was tossed from his longboards while en route to the disaster and landed back in the hospital with a "busted leg."

The weather cleared briefly, allowing races to be run at Howland Flat. Then the mountain furies reasserted themselves. The Keystone Mine and boarding house were then crushed by an avalanche, taking the lives of five men. A day later seven others were dug out still alive.

Amid all the carnage, racing tournaments continued. En route from Gibsonville to the races at Eureka Mine, two would-be spectators were overwhelmed by an avalanche in Slate Creek Canyon. The white shroud

Heavy winter came to Nevada City in 1910. During most winters there is not much snow at that elevation. Note trolley car. Elza J. Kilroy photo.

enveloped a miner named Jacob Black. His companion, Bill Eastland, burrowed up through the snow from where he had been buried and survived.

While Black's body was being dug out the races got under way and provided additional work for the volunteer rescue corps. The race track was only 15 feet wide where it crossed a ravine just before the finish poles. There, the officials had constructed a temporary bridge to the finish line. A rider named Matthew Webb met his Waterloo at the bridge when it proved too narrow a crossing for the eight men, running shoulder to shoulder. Webb suffered a broken leg when nudged off the righthand racing flank. He was sent cartwheeling into the ravine below.

Just how many men died in the snows of the Snowshoe Era will never be known. Newspapers during those years told of many tragic incidents similar to the following news item, taken from the Plumas National, Oct. 14, 1876: "The skeleton of a man was found in a bushy ravine, about four miles from Mohawk, by a party of hunters. A pair of snow-shoes was found near the body. It is supposed the man died during the heavy snow storms of last winter."

There were still living old-timers who in 1953 could recall the terrible six months which began in late November, 1889 and continued well into May, 1890. A continuing series of storms began that winter and by January 18, snow was piled up to record depths along the Yuba. There was 25 feet of snow on the level at Gold Lake; and 20 feet covering Mountain House and Forest City. Never in Downieville's history had there been such snow depths that encumbered travel about town. Through a hard pack of 18 feet, the citizens dug four tunnels. One stretched from Coster's butcher shop to Brown's hardware store; one ran from the livery stable to Blohn's saloon; another led from Wiggin's Place to the St. Charles Hotel, and another lined from Buzeinet's saloon to a mercantile store. Business, of course, came to a virtual standstill. There had been no communication for several days with mining camps in the higher elevations. From the Sacramento Valley, below, had come ominous reports indicative of what was yet to transpire that winter. Expressmen who snow-shoed into the Lost Sierra with copies of the Sacramento Union were not anxious to proceed after what they read in that newspaper. To be found in the Union of Jan. 6, and Jan. 13, 1890 were the following short items:

To the delight of Nevada County residents, the Nevada City Snow-shoe Band could trek anywhere in winter to offer a lively concert. The Nevada Daily Transcript, March 9, 1867 reported: "This excellent band arrived in town yesterday and will escort the Nevada Light Guard to the shooting grounds today. The boys came all the way from Washington to within 5 miles of Nevada City on snow-shoes. W.B.B. Collection.

"Malcolm McLeod, who undertook to carry, via snow-shoe, the mail from Nevada City to Washington in Nevada County was frozen to death, Jan. 5.

The following account of McLeod's demise can be found at the Western America SkiSport Museum. It appears in Robert I. Slyter's historical notes of the "Early-Washington, Nevada County Mining District."

"Malcolm F. McLeod and John Grissel left North Bloomfield at 2 p.m., Jan. 5, 1890 carrying 20 pounds of mail and express, each. They expected to make the trip to Washington in about five hours. Within two miles of Washington McLeod began to weaken.

"They reached Governor Thompson's cabin at the mouth of Jefferson Creek and wanted to stop there, but Thompson had neither food nor fuel. They had already discarded their snow-shoes. They continued to work their way through the snow, which was five or more feet deep on the trail.

"McLeod became so weak that Grissel had to drag or carry him. They got beyond Red Point and within a mile of Washington, when McLeod gave out completely.

"Grissel shouted for help. His cries were heard in the town. A party of 15 men, with a horse and Dr. Freeman, went out and found McLeod still breathing. He died at 4:30 a.m. as he was being carried into the McKee Hotel.

"McLeod, 28, was a native of California and employed as a mail carrier."

Slyter's report included the story of Mary Condon, an honor graduate of Washington's school, who was hired as a school teacher at the tiny nearby mining community of Omega. On April 6, 1882, when a severe snowstorm blocked all roads from her home at Moore's Flat to Omega, she donned her snow-shoes and arrived on time. The story circulated throughout Nevada County, making her a snow-shoe celebrity.

Winter tragedies in the mountains were a popular subject in the Sacramento press. Deaths were always front page news.

Another mail carrier perished further south near Yosemite, according to the Jan. 13, 1890, issue of the Sacramento Union.

"Alexander McGee, carrying the mail in Tulare County from Badger Pass to Moore and Smith's sawmill, and wearing snow-shoes, was frozen to death Jan. 13."

Others were to meet a similar fate that winter, including a snow-shoeing Chinese express carrier. The Celestial perished while packing provisions from Downieville to a mine at Gold Bluff. Not so unfortunate, however, was Carl Bracken, a snow-shoe expressman who made his headquarters at American Hill. He suffered a broken leg when caught in a snow slide but was found in time by a searching party and carried to Downieville.

Mining camp after mining camp was evacuated. Snow slides crashed down on all sides as miners snow-shoed to safety in Downieville. Here, too, avalanche conditions reigned. Slides struck the town. People arriving from the high country brought word that some miners had been left behind because there had not been enough snow-shoes for the use of all. Among those who had escaped were many who rode

Postmistress stands outside the tiny Post Office at Washington, destination of the ill-fated mailman, Malcolm McLeod. W.B.B. Collection.

tandem all the way behind another miner on snow-shoes.

Following destructive slides, which claimed several lives in Sierra City during March, 1890, fears were expressed for the safety of John Gatiker, a Grand Army member who lived alone near Bassett's Station. Led by the same John T. Mason, whose father had whipsawed snow-shoe billets at Rabbit Creek in the early years, a volunteer snow-shoe corps headed out to rescue the old soldier. It was a forlorn hope. Gatiker was found dead in the snow where he had been smothered by an avalanche that crashed from the peak above.

Well into May of that terrible year, death was still riding the snow-shoe trails. May 17, the Messenger carried the following item.

"DIES IN THE SNOW—Michael Ryan, en route from Forest City to his camp at Nelson's, died in the snow from cold. It is inferred that when he found he was lost, he concluded to lie down until morning and start out again. He had placed his snow-shoes against a tree, folded his hat and placed it under his head as a pillow. He undoubtedly died from cold and exhaustion. He was 60, a miner in these parts for many years. He was a well-informed man who could quote Shakespeare by the yard."

The folklore of the Lost Sierra is brimful with accounts of snow-shoe races won and lost against the Grim Reaper. A hair-breadth victory and two losing runs are expertly related by the old-timers.

One fine morning during the Winter of 1859, the snow-shoes of William Perkins, an old sailor, pointed down the trail toward Buck's Ranch. Sailor Perkins had cruised on many mighty leviathans of the deep, crossing icy seas before coming to the Feather's East Fork.

Stormy weather caught up with Perkins as he

Right, John Grissel, whose family owned a ranch in Washington, was the mail carrier who accompanied Malcolm McLeod on the ill-fated trip from Nevada City to Washington, Jan. 5, 1890.

Mary Condon, the snow-shoeing school teacher, friend of the Grissels who captivated Washingtonians when she snow-shoed nine miles to her Omega School, when deep snow covered the route. When her feat was published, she received several letters proposing marriage. W.B.B. Collection.

trudged toward Buck's. He was soon overcome by cold and fatigue, by the drowsy numbness that leads to death. Fenton B. Whiting and his assistant, while carrying a load of papers, packages and letters strapped to their backs, found Sailor Perkins unconscious in the snow. The pair had been fighting sturdily against the elements in an endeavor to make as quick time as possible to American Valley.

It was 10 p.m. when they stumbled over Sailor Perkins, who lay unconscious in the snow. Quickly raising him to his feet, they found he was unable to stand or move. He seemed to be beyond hope of resuscitation. Whiting dispatched his assistant to Buckeye to get help, blankets and restoratives. When these arrived, wine was forced down the throat of the benumbed old sailor. He began to show unmistakable signs of life and was carried through the snow to safety.

Not so fortunate were two other snow-shoe travelers in the same region during the winter of 1881. The resulting tragedy was also to claim one of their would-be rescuers. Losing out in the race with death were John Harold, a worker at the Monte Christo Mine and Mrs. John Nibecker, wife of the mining company's engineer. They both became lost and perished while on a trip to Spanish Peak. George Robinson, a top snow-shoer and member of the rescue mission, was crushed by an avalanche as he raced for help.

When the frozen bodies of the missing man and woman were found, the leader of the searching party designated Robinson, a 21-year-old youth and one of the region's greatest snow-shoe racing experts, to carry news of the tragedy to Meadow Valley, where a sleigh could be procured on which to remove the bodies. Robinson strapped on his snow-shoes and headed directly for a toll-house below. Later that day, when Robinson did not re-appear, other members of the rescue party descended the mountain. They found that young Robinson never had appeared. They immediately recalled seeing a large snowslide on the steepest portion of the mountainside. When they reached the slide area they found an immense mass of loose snow had slid across Robinson's trail. The story was written in the snow, where Robinson's snow-shoe tracks and the avalanche had joined. It was apparent that Robinson had come at full speed down the mountainside, riding a course diagonal to the avalanche. Robinson could not avoid the avalanche because the howling storm's swirling snow kept him from observing it.

Work was at once begun to find the body. Calls for assistance were spread throughout the nearby mines and soon almost 50 men had assembled. It took three day's work to uncover Robinson's body which was taken to his home, only a mile away. The funeral, two days later, was attended by a large gathering of snow-shoeists who had traveled many miles across the mountains to attend.

Challenges in the Lost Sierra often pitted man against other mountain dwellers sometimes more adept at surviving the deep snows. The folklore story of Sam Berry tells this best:

the story of sam berry

Samuel Berry was a quiet man of studious habits, according to the Lost Sierra story teller. Berry was seldom seen alone without a book, magazine or newspaper in his hands. He was from the north of Ireland, or Scottish parentage, and had come to America in his youth.

Sam Berry snow-shoed across Donner Summit in 1867. He was headed east and descended the granite cliffs of the famous pass, landing in Truckee. Sam sought a job as a ranch hand since the mines had no attraction for him. Sam's tracks in Truckee crossed those of Abram D. Church, a Sierra Valley ranch owner. In that manner, Sam came to Sierra Valley, where he met his end in a tangle with a grizzly bear.

Sam worked three summers on the Church ranch. While thus employed, he made many friends throughout the region. All the ranch owners and hands like friendly, intelligent Sam Berry. One of these friends was Sanford Morrison from Maine.

Sam took up a homestead with Morrison in a section known as Hamlin's Canyon, which debouched from the Lost Sierra and was full of fur-bearing game. During the snowy months the two men ran a trap line.

Early in the morning, Nov. 19, 1874, Sam kicked into his snow-shoe bindings and single poled up the canyon to collect trapped animals and rebait the traps. He never came back.

Well, 24 hours or so alone in the snow was not considered unusual in those days for a mountain man. It was not until the following day that Morrison became alarmed. He snow-shoed into the canyon. It was a fruitless search. The snow was deep, at least five feet, and with a storm threatening, Morrison returned to the valley to organize a large searching party. Fourteen men cruised the area on snow-shoes all the next day. They did not find a trace of Sam.

All of Sierra Valley then became aroused. On the third day there were 200 men scouring the hills. One of these men reported the discovery of what he believed to be be bear tracks. He reported that if they were bear tracks the animal was a monster.

Following the tracks, searchers came upon Sam's snow-shoes and his single pole, which proved to be close to the death scene. Evidently Sam had placed his snow-shoes against a tree and had been trudging about on foot when he and the bear came face to face. Sam ran and the bear followed in pursuit. It was a race through the snow without benefit of snow-shoes. Apparently Sam had fallen over a log and soon lost the contest.

The bear was supposed to have been one known as "Old Club-Foot." He was a grizzly that had been caught in a trap and lost several toes. In any event he had made short work of Sam Berry after catching him in a race through the snow.

References:

The Downieville Mountain Messenger
The Plumas National, Quincy
The Sacramento Union
Author's Interviews

Linking Nevada City and nearby Grass Valley was this suspension bridge. It linked the two cities and Auburn, California.

11

Man Killed At Saw pit

A difficulty occurred at Saw Pit Flat on last Saturday evening between Bob Francis and Bob Oliver, during which the latter was shot and killed by the former. Oliver called Francis a son of a b—h on the floor of a ball room. No further particulars have reached us."
DOWNIEVILLE MOUNTAIN MESSENGER, OCTOBER 3, 1868

There is perhaps no more involved a true story, typical of gold mining days, than that revolving around the killing of the world's first recognized ski champion by his bitterest racing rival.

An account of that affair, which saw Robert Oliver shot down by Robert Francis during a dance hall brawl, can be pieced together through folklore, court records, old newspaper accounts and the archives gathering dust in the Governor's office in Sacramento. These sources paint a picture of the snow-shoeing fraternity in a manner which might have been best reflected in the prose of such romanticists as Bret Harte or Joaquin Miller, had those early-day writers known about it.

The very atmosphere of Saw Pit Flat, a bustling mining camp in Onion Valley, late in the afternoon of Sept. 26, 1868, seemed to indicate something unusual was about to happen. Frock-coated men, their beards and locks newly trimmed, could be seen lounging around the only street of which the town could boast. Roughly dressed miners, just off shift from the Eagle Mine, trudged into town from the shaft up on the mountainside. Women circulated about. The usual dull routine of the camp was sparked because of the opening of a new saloon and dance hall by Saw Pit's hotel keeper, Bill Metcalf.

Metcalf, as the recognized snow-shoe racing dean of the community, had invited all his friends among the fast dope fraternity to enjoy the opening dance. The "boys" had been streaming into town all day long and those who brought wives or sweethearts with them were being doubly welcomed. Upon the faces of all the men could be observed a devil-may-care expression. Many already were congregated for a drink or two in the saloon and there was some talk of what might happen should both Bob Oliver and Bob Francis show up during the evening's doings. It was well known that only a year before, the two men had engaged in a fight during which Oliver had almost kicked Francis to death.

The attention of several miners gathered in front of Metcalf's place was drawn to a man walking into town. Bob Francis, a raw-boned and mustachioed miner of about 32 years of age, acknowledged greetings with a toss of his head. His high forehead and piercing eyes seemed to indicate that he could wield a powerful force for good or evil at the dance that evening.

Here are the details according to folklore, news clippings and a transcript of the trial:

Everyone had tried to calm troubled waters. Metcalf cried out:

"Hello, Bob. Will you come to the hall and take tickets for me?" "If I come," Francis replied proudly, "I'll buy my own way in."

A buzz of snow-shoe talk sprang up as Francis proceeded toward his cabin down the street. Soon fiddles were being tuned up. Those waiting for drinks to be served at the bar continued to use Francis and Oliver as the subject of conversation. Meanwhile Oliver, frock-coated and wearing his two-year-old championship belt, appeared on the scene. All eyes watched the wavy-haired and bearded snow-shoe dandy.

"Where was the fight you say?" asked one miner of Alex Kenn, an Onion Valley snow-shoeist who had been telling about the two racers' rivalry.

"It was in the old saloon," Kenn told his drinking companions. "It was a year ago last spring. Right soon after the races at La Porte. There was no snow on the ground, but it was before we washed up."

"Was it a fight with hands or weapons?"

"Boxing gloves, but it was a fight before they got through."

"Who got the best of their difficulty?"

"Well, Oliver did get the best I apprehend. I seen the fight. When I came in that house there was a full dozen people there. Francis wanted to put on the gloves to box. I seen Francis down and getting kicked."

"You saw Francis down and Oliver kicking him with a heavy pair of miner's boots?"

"Yes. That is the fight I saw there."

"Did somebody else take part in that fight before it got through?"

"Yes. Bill Clinch struck, or threw Francis on a table. I am not sure whether he struck him or not. I know he had him clinched. I could not tell whether he struck him. I told Clinch he had better leave him alone or I would take a hand in it. Francis could not stand up. He was cut somewhere on top of the head. Some of the boys took him down to the cabin and he never came out for two weeks."

A miner nodded to the barkeep, who looked down the line and asked, "Sugar or bitters, sirs?" But most seemed to prefer their liquor straight.

Kenn raised his glass and proposed:

"Here's to them. May both have heaps o' luck when the winter comes."

Out on the floor, Oliver was parading an attractive local widow, Eliza Kelly, through a complicated quadrille. The tall Cornishman and the lithe Irish girl made a striking couple. It was well known that Francis was also sweet on the girl and that his appearance at the dance might be a signal for another serious rupture between the two men. This possibility was under serious discussion as Oliver seated his partner and turned to approach the bar. At the same moment Francis strode in a door at the opposite end. The two rivals met face to face.

"Boys, be friends!" exclaimed Metcalf, grasping each by an elbow. "The drinks are on the house and you can pick partners for the next set."

Another set was formed when the liquor had been downed. The Kelly girl accepted Francis' proffered arm and whispered: "We don't want any trouble this evening."

Oliver's partner was Millie Delahunty, wife of a Gibsonville mine owner. Two other couples rounded out the quadrille. There was a silence in the hall like that which precedes a powder blast in a mining breast. Fiddles whined as watchers tensed.

The fourth time around the two adversaries collided, back to back. Francis, without looking over his shoulder, quit the floor, while the Kelly girl ran to a seat on the sidelines.

"Eliza, your partner's mad at you," a friend said.

Jumping to her feet, Eliza ran to Francis' side.

"Eliza, who did that?" he asked bitterly.

Francis looked so mad it frightened Eliza. She was doubly shaken because Oliver was approaching.

"I can lick any god-damned man who did that," Francis asserted as Oliver sat down beside her.

Francis then stood up and Eliza grasped his hand.

"Twas a woman who done it, Bob. If you want to fight anybody you will have to fight me."

Oliver looked up at Francis and laughed. There was a smile on his face. Francis, releasing himself from Eliza's grasp, wiped it off with a blow of his open hand. Friends moved in to grasp and restrain the two infuriated men.

"Ye son of a bitch," Oliver cried, "I have beat you once and I can beat you again."

"You Cornish son of a bitch," Francis taunted, "I'll cut your heart out." Friends led Oliver to the street, from where he shouted:

"Francis, if you have anything against me come out in the street if you want anything out of me."

"Keep quiet," a friend said, "this is no place for a

fuss or to have a fuss."

"I know it," Oliver admitted, "I would not care if I had not got my coat torn."

Inside the dance hall, things had quieted down. Francis had been persuaded to quit the place peaceably by a side door. Fifteen minutes passed and it was all right now for Oliver to go back in.

"There's not going to be any trouble," he told the crowd.

the killing

Framed in the opposite doorway was Francis. Something glittered as he raised his right hand. Women screamed; men dove beneath the bar.

"You son of a bitch," Francis cried as he stepped forth. The Colt in his hand boomed. Oliver fell back across a table, a lead bullet in his left breast. Francis made an effort to shoot again.

"No, no!" protested Francis as the weapon was twisted from his hand by Metcalf.

"Go home, Bob, go home," Metcalf admonished. "Wait for the law there. Oliver's hit bad, but may live through this."

The crowd was evenly divided among the friends of both men. While there were mutterings from those favorable to Oliver, none stepped forward to differ with Metcalf's advice. After all, the champion snow-shoeist might live to race again. If not, then some of the boys were ready to take care of things in their own way.

At daybreak, constable William Wood knew he must act swiftly to forestall the miners from taking the law into their own hands.

"He's not dead yet," Wood told Francis, "But we're going to take you down to the county seat. Things are kinda warm up around here for you."

"Bill, believe me," Francis said, "It was him or me. He was a cheat. It goes back to the snow-shoe races. He was a cheat. He started before the tap of the drum to win the belt. The son of a bitch. I would have killed him once before if I had had my pistol."

To reach the sanctuary of the Plumas County jail in Quincy, the two men avoided the usually traveled wagon road. (In 1867 Quincy became the county seat for La Porte and, of course, Saw Pit Flat, which is

Robert Francis

"Framed in the opposite doorway was Francis. Something glittered as he raised his right hand. Women screamed; men dove beneath the bar...."

much closer to Quincy).

"Hello, Bob," the jailer said, "how is this?"

"Oh, I shoot too straight."

Up in Saw Pit Flat the snow-shoe champion had just passed away.

Word of Oliver's death spread rapidly the length and breadth of Plumas and Sierra. In La Porte a special meeting of the Alturas Snow-shoe Club was set for Nov. 26. Placards announcing the gathering were posted throughout the hinterland. The killing, many feared, reflected some discredit on the snowshoeists. Now was the time to discuss the affair in an open and aboveboard manner.

The special meeting of the club, actually an "association of snowshoeists," by a vice-president, was called to order by Vice-President Charles W. Barnes of La Porte. He took charge in the absence of President John Conly. Representing other towns were John Clinch, Saw Pit; Abe Kleckner, Port Wine; George Winchell, Gibsonville; Hi Wallace, Howland Flat; N. B. Iseman, Grass Flat; Charles Scott, Poker Flat; D. Conlan, St. Louis; Doc Hall, Washington Hill; John Ward, Morristown; M. Tranor, S.T. Brewster and Alex H. Crew of La Porte. These officers represented the cream of the business, professional and mining world of the mountain communities. They were the leaders of organized snow-shoe racing. The Union Hotel banquet room was jammed with racers and club members to hear what they had to say.

Out of a discussion revolving about the known racing feud between the murdered champion and Francis, now

The Metcalf Hotel, scene of Robert Oliver shooting at Saw Pit Flat, Onion Valley, above, pictured about 1880. Eliza Kelly Metcalf, is shown center in front of her husband's hotel as she prepared to go snow-shoeing with two family members.

"...It was him or me. He was a cheat. It goes back to the snow-shoe races. He was a cheat...."

cooling his heels in the county jail, it was voted that:

"Hereafter a majority of the officers of the club shall settle all disputes in regard to the rights of the snow-shoe riders upon the track. A majority of the riders entered for each day's races shall decide all questions as to the locality of the track to be run upon and shall also determine the question as to the propriety of postponing the races when any objection is made to running on account of weather or state of the track."

By a roaring voice vote the snow-shoers then resolved:

"That the members of the Alturas Snow-shoe Club most deeply regret the unfortunate occurrence that took from our ranks by the hand of death one of our earliest and prominent members, Robert Oliver. That we feelingly sympathize with the bereaved relatives of our late friend in this hour of their affliction. That a copy of these resolutions be presented to the relatives of the deceased."

Speakers eulogized Oliver and recalled how the Cornishman had won the club's first championship belt in 1867 and almost repeated the following season. To win that race Oliver defeated John Gray Pollard, also of Saw Pit, after they had first run a dead heat in what should have been the final squad race. The decision had been protested by Francis, who himself had been eliminated by Oliver before the final squad was run.

After the meeting adjourned many of the out-of-town visitors spread about seeking liquid refreshments. The Saw Pit friends of the dead man gathered in Legan's Saloon, where a discussion of events which preceded the killing ensued.

Bill Wilson, a one-time co-worker of Francis, placed strategically at the bar, was answering questions for the boys.

"Bill, didn't you once work in the Eagle claim with Francis?" one of them asked.

"Yes, we were acquainted," Wilson replied, "We worked together."

"Did you ever talk with him in regards to Oliver?"

"Well, we talked about the La Porte snow-shoe races. That was last January when they was saying the races were coming off in a few weeks."

"Was Oliver as crazy as some say? What did Francis tell you?

"Well, I asked Francis if he was going to the snow-shoe races. He said he did not know whether he would go or not. I told him he had better stay to home and that he never would win anything if he went. I told him he had better work in place of going there—that the boys always beat him and that he could not beat Oliver. He said that it was best to stay home, but when the races came off, he went. I tells him again he won't win anything. And he says Oliver started before the tap of the drum—and that he had no right to the belt; that it belonged to Gray Pollard. He said that, 'If that son of a bitch ever speaks to me I will shut his wind off that he should never blow over me any more.'"

"You believed him when he said that?"

"I never thought anything of it. I did not know what he meant."

"He didn't say he was going to kill him, did he?"

"I didn't understand that he was going to kill him. I suppose he meant something. I never thought of it since, until that morning that Bob was shot."

"Did it not appear to you that it was revenge?"

"I did not think it amounted to anything at all until when Oliver was shot. Then I think maybe he had some old grudge."

There were angry murmurs along the bar.

As another round of liquor was served up, someone cried:

"We'll see that bastard hung."

A shout of approval echoed through the saloon.

But all the notable figures in the snow-shoeing world had not come into La Porte for the meeting that night. Even as Francis was being cursed in the town's saloons there were other men preparing a defense for him. These included Creed Haymond, the club's first president, a man then rapidly assuming legal stature throughout the state; Fenton B. Whiting, Quincy, who in later years wrote a county history; John Corbett, operator of a chain of mountain hotels and Fred

Saulter, a Gibsonville businessman and racer. There had been an unbreakable bond of friendship forged between these four man and Francis during the early pioneer days. All had been snow-shoe expressmen for Wells-Fargo and other companies operating through La Porte to Gibsonville, Table Rock and Poker Flat.

The friendship tie between Haymond and Francis was particularly close. Both were educated men. Haymond's background stemmed out of the law office of his father, W. C. Haymond, of Randolph County, Virginia. Francis had had a superior schooling in his native Canadian Province of Ontario. They both had come to the mines in 1852. They first met in Gibsonville. Each was about 16 years of age. First they mined and then they packed the mails and gold dust on snow-shoes. The two men became irrevocably linked through this association. While Francis was eventually convicted, it was his friendship with Haymond, that saved him from the noose.

the trial of bob francis

The "People vs. Robert Francis," was among the most vigorously contested trials ever held in Plumas County. The Canadian snow-shoer was indicted on April 6, 1869. A trial date was set for late May. The entire cast, judge, prosecutors, defense counsel, jurors and witnesses, was comprised of members of the Lost Sierra's snow-shoeing world.

Haymond, as chief defense counsel, tossed every conceivable legal obstacle out to delay the trial. When the case was called for a hearing, Haymond moved for a continuance in the absence of witnesses from Canada. He alleged they would testify that insanity was hereditary in the Francis family and extended back through three generations. This move failed to impress the presiding judge. Haymond then sought to temporarily halt the proceeding by claiming he was too ill to proceed.

The trial, however, got under way May 31, 1869.

Never had a more intense interest been manifested in a criminal action in Plumas County.

Snow-shoers flocked down in force from the mining camps, filling Quincy hotels and creating a problem for the court bailiff. The presiding jurist was District Judge Warren Key Sexton. District Attorney H. L. Geart, assisted by Judge Peter Van Clief, appeared for the prosecution. Haymond, John R. Buckbee, G. G. Clough and J. D. Goodwin stood on behalf of the prisoner.

The prosecution sought to prove a motive of revenge and premeditated murder revealed against a background of snow-shoe racing, while the main defense was a plea of insanity and the unwritten law of the mining camps: that it had been "one or the other" and Francis had got his man first after being called an epithet especially provoking to a Californian. As the trial concluded it was assumed by the prosecution and conceded by the defense that there must be either a verdict of murder in the first degree or of not guilty.

Apparently the jury was not entirely satisfied with the evidence, since it brought in a surprise verdict of murder in the second degree. The compromise, reached after a court adjournment of many hours, jolted both prosecution and defense. All had been resigned to seeing Francis hanged or set free. None had anticipated a guilty verdict in the lesser degree.

With Francis sentenced to 15 years, locked up in San Quentin, an appeal was made and lost in the California Supreme Court. However, Haymond was not one to give up easily when an old friend was involved. Strong appeals failed to secure a pardon from Gov. H. H. Haight.

Then, Dec. 8, 1871, came the inauguration of Newton Booth as chief executive of the Golden State. In a surprise move, the nineteenth session of the California Legislature, by a majority of both houses, recommended a pardon.

Here's the story as it was pieced together:

Francis, locked in his cell down by the Golden Gate, was in hopes that he did not have long to wait.

The tenth of April, 1872 began as just another day in jail for Francis. Ever since his confinement in San Quentin he had been discharging the duties of a clerk in the office of the captain of the guard. His conduct had been exemplary and it was believed that all of the prison officers were favorable to the pardon petition then pending before Governor Booth.

As Francis entered the office that morning, the captain said: "Sorry, Bob. There's nothing new from Sacramento today. Maybe the telegraph will have something later. We're all with you."

Francis assumed his accustomed place on the high-topped stool and thumbed idly through the captain's bookkeeping accounts. From the desk he could look through the barred windows across San Francisco Bay. Out there to the West somewhere was the Golden Gate, through which he had entered California en route to the gold fields 20 years earlier.

The onetime hardy snow-shoe runner was now sallow-faced from almost three years' confinement. Up in the mountains the racing season would just about be over and miners would be preparing to wash up as soon as there was sufficient water from melting snow.

Haymond and the others had not permitted Francis to testify in his own defense. Had that been a fatal mistake?

Alex Kenn had been his pal and come to his aid when Oliver and Clinch had all but kicked him to death. But it had been Kenn who brought the snow-shoe racing business into the trial testimony. Bill Wilson had offered telling testimony. The People's case had been bolstered by other witnesses such as Bill Metcalf and Eliza Kelly who were married two years ago. They had been called to testify for the prosecution, They had testified against him somewhat reluctantly.

However, Haymond and the boys had stood by him, including Fred Saulter, his pal from Gibsonville.

"Do you remember seeing Francis any time at La Porte in 1868?" Haymond asked.

"Yes, sir," Saulter came back. "I recollect seeing him here in February 1868."

"Did anything occur then to attract your attention?"

"Yes, sir. It was one evening we was there; it was the time of the snow-shoe races. I saw a man coming which I took to be Francis. I was always in the habit of calling him Bob.

"I says, 'Hello Bob, is that you?'

"I notice him. He came up, laid his hands on my shoulder. I asked him what was the matter? He said he had trouble. He said he was afraid Oliver would kill him. I was shocked with the conversation and thought I would turn it."

Haymond queried Saulter.

"What was his manner? Did he look as if he felt what he said?"

"He appeared to be very serious at the time of talking with me about it, and acted a little strange when I saw him. When I turned the conversation I told him I would guarantee that he would be safe," Saulter said.

"He met you in the street and put his hands on your shoulder and said that he was afraid that Oliver would kill him. There had been no conversation to allude to that, had there?"

"No, sir."

Saulter's cross-examination had been brief. He was followed to the stand by John Corbett.

"How long have you known Robert Francis?," Haymond boomed.

"About 13 years," Corbett replied, smiling down at his old friend.

"Where did you live when you first knew him?"

"I lived at Slate Creek House in Sierra County."

"How long did you continue to live in the vicinity where he lived?"

"The first period I lived near him was for eight years. Then I lived in Gibsonville, where he lived. He was a miner when I first knew him. He was in the express and gold dust business next. I was well acquainted with him."

"What were his habits as regards sociability.?"

"He was very sociable. He liked company. I believe he attended nearly every dance or social meeting parties."

"Did anything ever occur that called your attention especially to him?"

"Yes, sir. It was two years ago this month. The night of the primary election. He told me there was a pack or set of scoundrels at Saw Pit Flat that wanted to take his life; that he live in fear; that he believed that it would do them more good to kill him than anything else they could do. I thought it was strange and made light of it at the time. I asked for an explanation. He mentioned Oliver's name in connection with them as snow-shoe runners. Francis himself was a snow-shoe runner and used to be a member of the snow-shoe club at Gibsonville. There was a club there, but not duly organized as it was in other places. It was called the Gibsonville Snow-shoe Club."

Corbett testified how he paid him a visit in the Quincy jail three weeks before the trial began:

"He stated that he hoped that myself and family would not look upon him as a murderer; that he had done nothing only what a man had ought to do, and that he would do it again under similar circumstances. He said, `I wish the community at large entertained the same view that I do of the matter. John, it was just this way. I had to kill him or he would kill me. The only difference was, I was the quickest.'

"What did he state as the result of the trial?" Haymond queried.

"He seemed only to have one idea of that; he was going to be out of there in a short time and didn't like the idea of being confined in jail."

"Did he speak of any consciousness of being in any danger?"

"No. I expressed my surprise when I came out. His statements were not such as I expected, nor his manner. He treated the affair very lightly."

"Did you notice any figures on the wall outside of the jail?"

"Yes, sir."

Now Francis was chuckling to himself. "They were making me out to be crazy," he recalled, "and bringing in the law of the mining camps."

Fenton B. Whiting, his old Quincy friend, next came to mind. Whiting had testified before Corbett's cross-examination.

"Do you know who made those figures or pencil marks on the jail?"

"The defendant."

"What are those figures you say he made?"

"Horses, birds and various other things. He has drawn a picture of myself in the act of catching a Chinaman. I don't think he has done me justice."

The courtroom crowd had roared at that sally. Whiting stepped down and Corbett returned to the witness chair.

"At the time you saw Francis in Gibsonville," the prosecution asked, "and he told you that there was a set of fellows about Saw Pit Flat that wanted to kill him, did you ask him for his reasons?

"I did. He stated that they had made the attempt once. He was satisfied that they still entertained the intention to kill him. He mentioned Oliver as a snow-shoe runner."

"Are you sure that you understood Francis at the time. Was the conversation coherent?"

"I think he was a little excited."

"You have been a particular friend of his since you have been acquainted with him.?"

"Yes, sir."

"Taken an active part in his defense?"

"I don't know what may be called an active part. I felt an interest in it."

"Have you hunted up evidence for him?"

"No, sir. I have not."

Sitting in the guard room, Francis could remember the proceedings well. A doctor had told the court about a depression on the "right side of his head on the parietal bone." Francis searched for the place with a fingertip and flinched when it was found. "Maybe," he thought, "I was crazy then. God, if there'd only be some word from the governor."

It seemed like he could still hear Bill Metcalf testifying. Bill had not let him down. There was a man who could assume a neutral role. He had been the first champion in the old stake racing days.

"Did you ever see Francis and Oliver together prior to their difficulty?" the prosecution asked Metcalf.

"Yes. In the barroom the night of this fuss. Both were standing up to the counter, taking a drink, both together. It was before the fuss in the ballroom."

"Did you see them together at the La Porte races?"

"I have known of their being there."

"Did you see them both there at those races?"

"Yes, sir. I have seen them both there."

"Did they both run in the race?"

"I know they were both there running snow-shoes."

"Do you know whether Francis was in the habit of drinking intoxicating liquor?"

"I have seen him drink once in a while."

"How many times did you see him drink the night of the difficulty?"

"Only twice, or three times, so far as I seen."

"Did you see anything unusual or strange in his manner at that time?"

"I did not."

Francis was now thinking back to the final moments of the courtroom scene that day. The plea to the jury in his behalf had elicited the highest admiration of the bench, bar and auditors. He had been elated momentarily, only to have his hopes dashed when the jury foreman handed the court a slip: "We, the jury, in the case of the People versus Robert Francis find the said Robert Francis guilty of murder in the second degree."

Powerful influences had been at work during the past two years seeking his pardon. Even now in Sacramento the matter was under consideration of Governor Booth. But to Francis there was only a hope. Ex-Governor Henry Haight had declined to act even though some of California's most prominent citizens had joined in the plea for clemency. Haymond was leading the effort to secure a pardon, his plea stating: "I have known Mr. Francis for many years as a man of standing and character in the community in which he lived. I believe the act was than of an insane mind and that he ought not to have been convicted. I cheerfully join in the appeal for executive clemency and earnestly recommend the full pardon of the defendant."

But Governor Haight had been unmoved, and then stood like a stone wall against a final plea written by Francis himself, from his San Quentin cell, Feb. 22, 1871.

The plea stated: "It is useless for me to recapitulate the merits or demerits of my case, as that has already been done by my kind friend, Mr. Haymond. I have only to say that I have been unjustly dealt with, and suffered to remain in prison, for a crime of which I am not guilty.

"As an R. A. Mason, I now make my last appeal to your excellency, to release me from my present position. Accompanying this appeal is my mask as a surety. This boon I ask at your hand as an R. A. Mason, feeling conscious that my appeal is not to be in vain. I make this appeal to your Excellency not because I think Masons are bound to assist an undeserving Brother, but because I know I am not unworthy. Did I feel that I was guilty of the crime for which I am now suffering, I would willingly submit, rather than bring discredit on the order by calling myself a Mason.

"Hoping your Excellency will give my case your consideration and hasten my release from the confinement, I am your Excellency's Humble Servant."

On that dull April day as Francis sat in prison, the High Sierra seemed years and miles behind. But the affairs of the onetime snow-shoe runner were rushing to a climax.

the pardon

The telegraph was chattering between Sacramento and San Francisco. An operator, in copper-plate handwriting, took down a message for the prison warden:

> "Whereas, at the May term, A.D. 1869, of the Second Judicial District Court held in and for the County of Plumas in said state, Robert Francis was tried and convicted of the crime of murder in the second degree and sentenced to undergo an imprisonment in the State Prison for a term of 15 years;
>
> "And whereas, the Legislature of the State of California at its Nineteenth Session, by a majority vote of both Houses recommended his pardon, and a careful examination of the testimony taken at the trial which resulted in his conviction shows the Legislative recommendation to have been well founded,
>
> Now, therefore, by virtue of the authority in me vested by the Constitution and the laws of this State, I Newton Booth, hereby pardon the said Robert Francis, and order on the receipt of these presents that he be discharged from custody, and restored to all the rights and privileges of citizenship to which he was entitled before his aforesaid conviction and imprisonment.
>
> Witness my hand and the great seal of State, at office, in the City of Sacramento, California, this tenth day of April in the year of our Lord one thousand eight hundred and seventy-two. Newton Booth, Governor."

The message was delivered to the warden's office as dusk closed in around the prison walls. Francis was still at his desk in the captain's office when the good news was relayed.

"What do you intend to do, Bob, when you get out of here?" the friendly guards queried as they pounded his shoulders and shook his hand.

"I don't know. I'll have to make up my mind whether to go to the mines or back to Canada."

"What in God's name would you go to Canada for?" they asked.

"I am going back to get married. I ought to have done that years ago."

References:

Transcript of Bob Francis trial provided by former Secretary of State, Frank Jordan.
Downieville Mountain Messenger.
Archives in the Governor's office.
Interviews and letters from Plumas County residents.
The Sacramento Union.

From the files of Frank Jordon, former California Secretary of State.

12

The Feather

> *Out looking for what we called "old-time skiers," it was Birdie Swingle, Quincy, who told us that if we would only ask for "snow-shoers" she could put us on the right track...The author.*

new roads to the lost sierra

The Feather River provides two gateways to the Lost Sierra, one from the north and one from the east.

The Lost Sierra's, north entrance is reached at a turnoff from a point two miles east of Quincy in American Valley. The tortuous dirt road is what long ago was described as the "finest mountain road in the state." Running over a rugged, twisting and narrow 35 mile stretch, the road climbs lofty ridges, drops through Pilot Peak which dominates Onion Valley, then winds through dense forest to the snow-shoe country capital, La Porte.

The Lost Sierra's eastern gateway is reached by turning south at Blairsden, 18 miles farther east up the Feather River Highway, US Route 70, from Quincy. From that turnoff it is only six miles. The route passes through Mohawk, via the old Emigrant Trail, then enters picturesque Johnsville, headquarters of Plumas-Eureka State Park. That's where the more adventurous traveler can, with great care, advance over a primitive

In this circa 1954 photo Birdie Swingle recalled how a record was made in 1873 by Tommy Todd who traveled 1804 feet in 14 seconds (87.857 mph). Only Tommy knew what had gone into the dope he used.

Birdie also recalled, "I was asked to write an article on Plumas County (history) not to exceed 300 words. "Who?" she asked, "... has lived here for 75 years? Whose grandfather came in 1852, ... the man that raised the Madagascar roosters used to stage the cockfights in Virginia City during the gold rush days? Who could tell it in 300 words?" Plumas County Historical Museum photo.

road directly into the Lost Sierra.

The era of travel no longer dependent on snowshoes, such as roads for pack trails and stages for saddle-trains, began early. Emigrants in 1851 hacked a trail through the mountain wilderness from Johnsville to Gibsonville. Its stupendous grades later were improved and remain today as a breath-taking monument to the transit and levels of Quicksilver Charley Hendel, found in an earlier chapter.

Not until 1865, however, did widespread agitation focus Plumas County citizens' demand for a roadway linking Quincy to the higher elevation mining camps. At that time the American Valley ranchers sorely felt the need of a market for their products. The local demand was trifling compared with the supply, but beyond the ridges to the south, residents of Saw Pit, Onion Valley, Gibsonville, La Porte, Whiskey Diggings and other mining camps were annually consuming large amounts of farm and dairy products. They could use more if they could get them delivered.

Construction of the road, authorized by the Plumas County electorate in 1866, was completed the following year. The building of the road was a pet project of the people of La Porte. It was financed through John Conly and Alex Crew, Rabbit Creek banking tycoons. Both were snow-shoers who held official Alturas Snow-shoe Club titles. The snow-shoers of La Porte had just succeeded in getting their community cut loose from Sierra County and annexed by Plumas County. When the road was opened, a gala celebration was to be staged at La Porte. Residents wanted the world to know their community was in fact, as well as in name, a part of Plumas County.

Over the newly completed roadway rolled carriages and stage coaches carrying ranchers and county officials to the festivities, American Valley lies some 3,300 feet above sea level. Celebrants found comparatively easy going over the first 12 miles to Nelson Point. From there it was up, up and up, soaring some 3500 feet in the next five miles. The wagons bumped over a seven-foot wide roadbed gashed out of a steep mountainside. Then they crossed over ridge tops, snaked through Onion Valley and Pilot Peak and entered Gibsonville. There the route threaded through thick pine forest toward the excitement on the banks of Rabbit Creek.

Johnsville snowed under. The town where Birdie Swingle taught school in 1896. Her first real adventures in the snow began in Johnsville. She soon was claimed as a bride by Jerry Curtis, a mining engineer, who in 1917 was shot and killed by a crazed miner. The little widow married again in 1924, tying the knot with Andy Swingle, a veteran whip of the stage coach days. Photo: Lynn Douglas Collection.

Today outdoor recreation has become the life blood of high mountain territories such as abound in Plumas County. Mining activity has become negligible during the passing years. Timber production has fallen off, too, though it remains the principal money crop. As a result, the spectacular Quincy-La Porte road to the Lost Sierra and the relics of the Johnsville area seemed destined to attract only those interested in historic and scenic attractions. The back country beckons to an adventurous few of the thousands of tourists who each season trek through the Feather River Canyon.

Looking back, it must be remembered that the Western Pacific's steel was not snaked through the

Feather River country until 1912. Further, the present U.S. Route 70, the state's lowest trans-Sierra pass, was not completed until the mid-thirties. That these routes of travel skirted the snow-shoer's fabulous Lost Sierra was overlooked by promotional-minded authorities, even though those old-timers, who had been born and raised in the Feather River country, tried to have such historical facts emphasized in the area's tourist literature.

oldtimers' recollections

Mrs. Andrew (Birdie) Swingle of Quincy and her fellow members of the History and Landmarks Committee of the Plumas Pioneers' Parlor No. 219, Native Daughters of the Golden West, were determined to put the Lost Sierra on the map.

During the 1940's Mrs. Swingle and her committee members had been both incensed and amused by the scant recognition given snow-shoeing history. They were not only critical of the area's promotion types, but also took exception to those who had written about ski history. All during that period they had been digging up documented historical facts about the Snow-shoe Era. They planned to someday place their treasured information in the archives of Plumas County.

They succeeded, since their collected work is now available at Plumas County Museum in Quincy.

Mrs. Swingle, known in her youth as Anna Birdena Haun, was included among the more famous snow-shoeists at the turn of the century. Everyone called her "Birdie" and the fact that she had arrived on March 22, 1875, a balmy springlike day, could have been the reason Birdena was shortened into her time-honored sobriquet. The birth of Birdie launched a very busy day in Quincy's justly famous American Ranch Hotel, as travelers found their way to its newly opened doors. It had been the first sawed-lumber structure in American Valley and the birthplace of Plumas County authority in 1854.

Dainty little Birdie Haun grew up in the town of Quincy. She attended local schools, wherein in those days, of necessity, part of the curriculum was learning how to snow-shoe. So it was only natural that later in her life, when Birdie was certified to teach school, she should be sent to remote Nelson Point to pass along her knowledge to the younger generation. That was in 1896. It follows that one memorable day, in that wilderness, Birdie also learned to ride tandem, as we shall see.

From Nelson Point Birdie transferred to Mohawk, then to nearby Johnsville. That town now boasts Plumas-Eureka State Park. The ghost town stands at the eastern entrance to the Lost Sierra. It was at

Lenny O'Rourke, left, is pictured "yarning" with modern-day skier Philip Miller at Johnsville in 1938. WBB photo.

Johnsville that a mining expert from the mountainous snow country found and soon claimed the little school teacher for his bride.

Jerry Curtis was a noted mining engineer. He married Birdie during the summer of 1904. The previous winter had seen Jerry displaying lots of snow-shoe prowess in the neighborhood of Birdie's Johnsville school house. He was employed at the nearby Four Hills mine. One of his more specific chores was to slide down the mountainside, loaded with gold bricks for shipment to San Francisco.

As a bride Birdie was taken by her Jerry to reside at the Gold Valley Mine in the very heart of the snow country. From the mine the snow-shoe runs stretched in all directions.

Birdie, a gracious little lady who had been a bride 46 years prior to our interview, recalled those days with relish. She told how riding tandem behind her husband had been a new adventure.

"It was early winter when the first snows came," she related. " Mister Curtis made himself a 13-foot pair of snow-shoes with two sets of toe straps. He was an expert, and I rode behind him. When the lakes above Johnsville were frozen we rode across them in a straight downhill stretch. I grasped Mister Curtis around the waist and we sped swiftly along while he guided or braked with his single pole.

"Once on a trip to Downieville we counted places where seven snowslides had come down the mountainside. Mister Curtis always seemed to have some instinct that told him when it was safe for us to make the trip. We'd leave the mine and he would make his tandem snow-shoes go fast when we were below dangerous-looking slopes.

"We lived in a two-story log cabin at the Gold Valley Mine. In a hard winter we snow-shoed out from the second story windows. The cabin was so snug and warm that I raised blooming geraniums all winter. For two years I was the only woman in the mining camp, but we never felt lonely or outcast living up in the snows."

The couple's snowy idyl was ended abruptly in 1917 when an insane miner shot and killed Curtis. The shooting came during an argument in an isolated mining area.

Seven years later the little widow married again. Her second husband was Andy Swingle, a veteran whip of the stage coach days. When I talked to them, Birdie and 84-year-old Andy were living in Quincy where I had first met the historically minded little lady

Out looking for what I called "old-time skiers," it was Mrs. Swingle who told me that if I would only ask for "snow-shoers" she could put me on the right track. Through her I met Lenny O'Rourke, an old race rider and dopist of local fame at Mohawk. Through O'Rourke the gates of the Johnsville area legends were opened for me into the Lost Sierra snows beyond.

In those days Lenny O'Rourke more often than not was referred to as "one of those Plumas prevaricators." The long snow-shoes he made and displayed each summer were referred to by the city slickers as the "product of an imaginative mind." Lenny, a twinkly little guy, was wont to awe visitors who visited his hinterland by pointing out Eureka Peak, jutting several thousand feet above the valley floor. Assured of their attention he would calmly say:

"That's the mountain on the California State Seal. There, down along the ridge, is our race track," he told them.

Johnsville, pictured Jan 20, 1913. A lone skier is seen carrying his snow-shoes. Eureka Peak looms high in the background. Lenny O'Rourke, former Plumas County chamber chief, always asserted that Eureka Peak, "is the mountain stamped on the California State Seal." Photo: O'Rourke Family collection

"Race track!" the visitor usually exclaimed, adding, "Now just because I happen to be from the big city, don't try and kid me that you can run horses up there."

"Horses? Gosh no!" Lenny would respond. He'd put on a roguish look and say, "That's our snow-shoe race track. Don't you believe anything about this country at all?

"Today we haven't many people living in these parts, but back in the boom years there were several thousand. Every weekend during the winter months the miners would come from miles around and run races for purses as high as $1,500. The amount of money changing hands in side bets was really unbelievable. It all started during the Gold Rush days."

Then he'd look very wise and say, "There are some who say that, way back in the hills, it's still going on to this day."

At this point the not-so-gullible visitor usually found it convenient to fumble with his fishing tackle and, not daring to look Lenny in the eye, assert: "Well, it's all news to me. What kind of snow-shoes did you really use?"

"Our snow-shoes were what you dudes call skis," Lenny would cheerfully explain. "They were eight to twelve feet long and plenty fast when all doped up. We used only one pole. It was about six feet long and two inches thick. It came in handy for balancing and when used as a brake."

Properly subdued, the vacationer would hurry back to his friends at the Feather River Inn, where there would be many chuckles over the tales as told by the Plumas County boys.

Documentation of the beginnings of gold mining beneath Eureka Peak commenced during the summer of 1851, the same season that emigrants skirted through the region and hacked a road through to Gibsonville.

Folklore of the old-timers in the nineteen-thirties had some uncommon tales such as that Chilean prospectors struck it rich near Johnsville in 1845, three years before the widely heralded discovery at Sutter's Fort brought on the rush of the Forty-niners.

Johnsville, founded in 1876, is now a unique year-around resort at the foot of Eureka Peak. It still retains many colorful characteristics of the Snow-shoe Era.

Much historical material can be found in Johnsville's Plumas-Eureka State Park Museum. Prior to that time the principal communities were Jamison City and the "City of 76."

The first mining locations were posted by a group of nine prospectors. "The Original Nine," they called themselves in May of 1851. Satisfied that they, indeed, had "found it," they named their ledge and mining company: "Eureka." The news soon spread and miners began to pour in from the Middle Feather. The newcomers also organized another company and because of the fact that they numbered just seventy-six souls, they bestowed the name of "Washington" on their location and then laid out a townsite on Jamison Creek and, naturally, called it the "City of '76."

The mines on Eureka Peak, which originally had been called Gold Mountain by passing emigrants, were of the quartz variety. Soon all was hustle and bustle. Ore from the gold-bearing ledges was packed down the mountainside to be crushed fine in primitive mule-powered grinders called arrastras. The ore was washed out with the bountiful supply of water which blessed the region. History does not explain how it was that the miners were able to set up arrastras in such a speedy fashion, but regional folklore credits the introduction of such mining machinery to the Chilean miners who had, apparently, come six years earlier. It appears the American miners helped themselves, when they got there, to what they found had been used by the legendary Chileans.

When we first visited Johnsville in 1931, relics of its vibrant past were plentiful. The town welcomed tourists as an historic point of interest. The site had been turned into California's Plumas-Eureka Park.

When we came back a few years later we met Steve Pezzola, "The Mayor of Johnsville".

Pezzola was then a 70-year-old snow-shoe maker and owner of a nearby placer claim which had liberally blessed him with the better things of life. Steve's edicts were not backed with the authority of any laws, but his word was good enough to awe the summertime visitors and many of the permanent residents who paid him attention.

Steve's "mayorality" was bestowed on him in 1938 by Governor Merriam during a convention of the

The Plumas Argus, in 1859 ran the March 26 letter from Poorman's Creek mining camp, two miles from then booming Onion Valley on the Quincy - La Porte Road. This is the first newspaper account of a Lost Sierra community on snowshoes, which states: "Everybody's on skis...a person can run over the light and new fallen snow at railroad speed."

Northeastern Section of the California Newspaper Publishers' Association at the Canyon Inn.

The honor came as a reward for Pezzola's resourceful rescue of the California's chief executive from one heck of predicament.

Flash floods turned loose by melting snows had cut the water line leading to the Canyon Inn. Steve rushed completion of an emergency pipeline so the governor and newspaper publishers could sluice off an accumulation of mountain dust.

Well refreshed from the cold waters rushing down from the melting snows above, Governor Merriam acknowledged his debt of gratitude by calling Steve before the assemblage and declared, "I name you mayor of Johnsville."

Further, in those days, Steve enjoyed the reputation of sage. The region, which in its heyday had a population of four or five thousand persons between Mohawk and Nelson Creek, included many mining camps thereabouts. Pezzola was the son of pioneer parents, and had a keen recollection of the folklore they handed down to him. His personal observations provided secondary documentation. He firmly believed that Chilean miners were the very first to work the Eureka quartz because that was what original old-timers told him when he was a youngster.

"When I was a young men," he said,"everyone always talked about the Chileans.

"Then there was Marakeet," he said gravely.

Marakeet, it seems, was a Chilean voluptuary who from the very early days operated love-for-pay establishments throughout the area.

Marakeet, as the miners called her, or Marachita, as she preferred it, was to become famous. It was not so much for her practice of the world's oldest profession as it was for her possession of a remarkable curative herb known as the "Devil's Shoestring."

The folklore about Marakeet and the Chilean miners first came to my attention in 1954. To refresh my memory I had a chat with "Mayor" Steve.

Steve knew the story of Marakeet well.

"Marakeet was a hurdy gurdy girl, you bet. But she became well thought of in her later years. When she died in the late nineties everyone in Johnsville turned out to honor her. That was especially true of the Pezzola family. Our family owed her a debt of gratitude.

"A member of our family had a cancer of the breast. After all the doctors in these parts had given her up, she went to Marakeet. It's hard to believe, but Marakeet had an herb known as the 'Devil's Shoestring.'

With it the cure was done.

"That was way back in the eighties, when I was just a young boy, but that member of our family lived to die of old age almost 40 years later. Of course, our family always believed good of Marakeet after that. She and my mother became great friends. That's how I truly know the whole story," he said.

I asked about the legend of the Chilean miners.

"Did Chilean miners actually come here in 1845?" he answered. "Well, they probably did. Anyway, Marakeet told my mother that she left Chile for California several years before the "big rush." She told my mother that some of her countrymen had brought her across the mountains on muleback. Later she was joined by several Spanish girls.

Steve Pezzola, "Honorary' Mayor of Johnsville is shown in 1938 inspecting the fire equipment at the Johnsville fire house, while the fire house dogs inspect him.

"According to my mother, in 1851 Marakeet was running a dancehall and drinking place alongside the Emigrant Trail.

"I have always wondered if the miners had snow-shoes that very first winter. Johnsville, of course, is in the deep snow country, but only six miles from the true snow line. Because it was easy to get to lower altitudes, the early miners were not scared out of the mountains when the snow came. Provisions could be packed in from Beckwourth, and whether the men had snow-shoes or not did not matter. Somehow they managed to travel about. Many of their trails led to Marakeet's place.

"That March six of those miners and all the girls except Marakeet, were killed in a snowslide.

"Marakeet and her girls entertained in a log cabin. Underneath the cabin was a storage space for the liquor supply. Maybe she made her own down there. Anyway, on that winter night the carousing miners demanded more to drink. It was those demands that saved Marakeet's life. She went down below just before the avalanche struck. She was dug out alive two days later," Steve said.

The legend of Marakeet's bordello in the snow is one of the very best stories told by the Plumas County boys. She continued as a madam for another 45 years and it is true that the miners did snow-shoe down from the mines to see the girls. In fact, "Mayor" Steve recalls that they used to pass him on the way down while he was going up to the Four Hills mine.

Young Steve, in the early days, was a snow-shoe packer who packed in flasks of mercury. This quicksilver's use was strictly for the gold amalgamating tables. Steve pointed out: "There's no joke in that, even though all the snow-shoe tracks passed Marakeet's place."

Snow-shoe racing early became a mania for the Eureka Peak miners. During the sixties and seventies there was a race track below Jamison City. In later years the scene of activities was moved to the upper and speedier slopes of Eureka. There, during the nineties, many spectacular tournaments were held and several notable records were set.

Because Jamison City tournaments were readily accessible to outsiders, snow-shoe races there soon attracted spectators from the lower regions.

Onlookers soon encroached upon the officials at the snow field's finish line. To eliminate this interference special stands were constructed for the February 1872 races. From this vantage point the races were viewed by visiting newspaper editors. Among them was a writer for the Plumas National, published in Quincy.

The National reporter wrote:

"George E. Cook, race judge, while running down the track, fell and dislocated a knee. Four or five of the boys, who hastened to his relief, took hold readily, and by a judicious pull got it into place again, making a snap like a pistol shot!

"The course was 1676 feet. Starting poles were up on top of the hill, where the ground was nearly level; at about 30 feet past the poles the track turned down to a grade of thirty degrees, then ran across a grade of sixteen degrees to the turn of the second pitch, which had the same

George Woodward, millwright and snow-shoemaker, came to California during the Gold Rush. He settled in Johnsville, where he supervised construction of the Plumas-Eureka Mine 40-stamp mill. The Columbus, Ohio, State Journal, April 2, 1849, listed George Woodward in a group headed west under Capt. Joseph Hunter. It was headlined: "The Boys Are Off For California" and reported: "Two companies of adventurers, of 30 each, residents of Columbus, have this day taken leave of their fellow citizens and started for the El Dorado of the Sacramento Valley." Photo: Woodward family genealogy

slope degree as the upper 300 feet; then came a graceful sweep to a flat of about seven degrees to the finish poles and beyond."

A miner named Louis Christopher was the big money winner, but a neck-and-neck squad duel between a couple of men named Charles Hanson and Charles McDonald won most of the newspaper space. Running shoulder to shoulder at 80 miles an hour in what today would be called a photo finish, the two snow-shoe jockeys took to their poles.

Hanson made "a pass at McDonald, who in turn threw his pole at Hanson, after being hit on the jaw." The two men then "went through the poles somewhat like a Catherine wheel on a Fourth of July celebration."

McDonald was declared the winner, and officials made him and Hanson shake hands to make up.

the emerging champions

The regional racing classics reached new heights when the Johnsville boys took over in the mid-eighties to stage tournaments on the upper snow field reaches of Eureka Peak. Squads then were run over tracks 1800 to 2,500 feet in length and many match races one and two miles long were contested. At the nearby Plumas-Eureka Mine there was an ore bucket tramway used by the snow-shoers. It was the world's very first ski lift old-timers will tell you.

Among the great riders of this later era were the Woodward boys and girls, the progeny of George Woodward, who had been among those establishing the "City of 76."

They ruled the local speed roster until the turn of the century. All told there were eleven Woodward youngsters, the more famous of snow-shoers were Frank, John, Arthur, Florence and Alice. Frank in his day was the accredited local champion, many times having bettered 100 feet per second from a standing start over the measured distances raced.

Old man Woodward was Johnsville's most noted snow-shoe manufacturer. He came by that ability naturally. A native of Wilmington, Delaware, he had been apprenticed as a carpenter in his youth. He could "throw-iron a spruce billet with the most expert."

Snow-shoeing in the days when George and Martha Woodward arrived in Eureka, Plumas County, can be best recalled in the words of their daughter, Fannie Woodward Hunsinger, in her Autobiography, "Counting My Blessings," handwritten in 1934-35 solely for her family:

Sketch of the Plumas-Eureka Mine, "Grandfather was a 49er." from the Woodward family genealogy.

"I was born in Mohawk Valley, Plumas County, California, one-half mile from where the Mohawk Hotel stands, on November 4, 1863.

"I can almost smell the sweet spring odor of the fresh earth under the pines, where the snow went off first. And what joy it was to find the first violet or other flower!

"Then for winter fun, when the snow was on the ground and crusted over, we used to take a large milk pan or dish pan, or even a shovel, and go sliding and whirling down the hill. What fun it was, but it was really dangerous although I have no recollection of any of us ever getting hurt. It is mostly the pleasant memories that linger in my mind. Then there was the sleighing behind prancing horses and jingling sleighbells, in neat cutters or home-made sleighs. In those days families visited each other. The whole family would go and spend the day at a neighbors, and the children and grownups of both families would thoroughly enjoy the visit. One of our neighbors at the extreme east end of the valley, had the finest cutter, the most spirited team and the most musical bells. We could always tell when Mr. McLear was coming in his sleigh by the sweet toned bells and many delightful rides have we enjoyed with him. A thrill we do not get nowadays in automobiles, although we would not like to go back to the slower mode of horse travel.

"I can remember on Christmas, in that old home by the mill, I got a little wooly sheep on a stand.

"So many incidents of my childhood and young girlhood come crowding into my thoughts that it would make more than a book to write them all down. Our hats didn't fit over the head like they do now, but set pretty well on top of our heads. We children wore elastic bands from the hats under our chins to hold them on. The ladies wore the elastic under their hair, which was worn coiled low at the neck. Our underclothes were muslin chemise, canton flannel bloomers made with bands to button just below the knees, and always two skirts. In summer, instead of the bloomers, we wore panties made from floursacks, trimmed in crocheted lace. The ruffled pantalets to the ankles were before my time. Our nightgowns were also made from floursacks. Flour always came in 100 lb. sacks of good quality, and as all baking of every kind had to be done at home, it took a great deal of flour for a large family. Our underclothes were always trimmed with homemade lace. White skirts, with three-yard ruffles around the bottom, were trimmed with the crochet edging. I cannot remember when I didn't crochet. No lace to be bought in those days and the lace was made just from common spool thread, J.P. Coats, which we still buy.

Augustus Cassius (Cass) & Fannie (Birdie) Hunsinger, at their wedding, Jan 31, 1883. Woodward genealogy.

"When I was about seven years old my father went to the Eureka Mine five miles away where he helped build a 40 stamp quartz mill and other buildings. He was a carpenter and millwright.

"After we moved to Eureka, we didn't use an old pan for sliding downhill on the snow, for everybody had skis. Snow-shoes we called them then, although they were nothing like the webbed snowshoes of today and not just like skis either. The ladies and girls' snow-shoes were eight feet long, the men's 10 feet. Racers were longer. We rode our shoes to school and everywhere we went, and oh, what real pleasure we had sliding down the hills and mountains in the keen, crisp air. The school house faced south. The snow, which fell 10 feet on a level and 30 or 40 feet in drifts, blew away from the south side of the buildings and piled up at the northside. This made a nice little hill at the north side, and during recess three or four of us would get behind one of the boys, all standing up on his snow-shoes with hands on the hips of the one in front of us, and away we'd go, over the jump. Of course it was nothing like professionals jump now, but it was quite a jump, and I assure you it was thrilling. The deeper the snow fell, the better we liked it, as it covered all the brush and made smooth riding. I've often wondered in later years if our mothers liked it to fall deep as well as we children did, but I don't remember ever hearing a complaint from any of the grownups. Young men, instead of taking their girls automobile or buggy riding, took them snow-shoeing in winter and for a walk in the summer. It was not so expensive, but just as thrilling.

"We wore plenty of clothing those winter days, with leggings and over-shoes—arctics, they were called, also thick green veils. Now I wonder how we ever breathed through those veils, but they protected our eyes and faces from the sun on the snow. Sometimes we would get snow-blind, which was not at all nice. We would have to stay in bed with either a poultice of cold tea leaves or grated potatoes over our eyes, for sometimes days at a time. It is not fun being snow-blind for the eyes burn like fire.

"The mines used to close down several times each winter to allow the workers to watch and many to participate in the snow-shoe races, which were the big event of the winter, and big money was put up for the winners. My brother Frank and some cousins were among the good riders. Even babies, two and three years old, had their tiny snow-shoes and learned to use them in the house. We always used a pole with a button on the end. It helped us on a level or going up hill and served as a brake going down. We had moccasins made from bed ticking to fit over the back of the shoes and tie around the heel piece. With these on we could walk on our shoes going up hill, as the moccasins prevented them from slipping.

"I remember one year it began to snow on April First and snowed until the 16th. As usual, the snow piled up at the north end of the house, and Frank made a tunnel from the upstairs window on that side.

Johnsville Schoolhouse, March 2, 1914, pictures youngsters lined up on their snow-shoes ready for fun on the snow. Lynn Douglas Collection.

"It never seemed to get so cold there as it does here. The snow banked around the house seemed to keep it warmer, and yet, when I think of it, the water would be frozen in the pipe for quite a while and we would melt snow for water. A door opened onto the porch at the northside of the kitchen and the snow would pile above the roof of the door, so all we had to do was open that door and scoop the snow up with a ladle or dipper, into a clean kettle or boiler, then melt it on the stove. We had to keep adding snow as it melted, for it takes a lot of snow to make a kettle of water. The snow used to get crusted over so solid that they drove their beef cattle up from the valley on top of it. They had their own slaughter house and did their own butchering there at Eureka."

Snow-shoe racers, posed in full speed stance, on Lexington Hill, La Porte, 1906; from left, Buck Mullen, Mike Bustillos, Joe Mullen and Joe Bustillos. Photo: Plumas County Museum.

Sons Arthur and John Woodward followed in their father's footsteps and as late as the mid-thirties were experimenting with new models and dopes in Susanville on the Sierra's eastern slope. In those years Arthur told this writer that,"boughten skis haven't the speed of snow-shoes; we know because we have tried them out."

Further, the brothers studied modern ski techniques in comparison to the manner in which they rode once upon a time. Arthur explained that the snow-shoe riders were at full speed almost from the start due to the exceptional quality of their racing dopes. He said they leaned forward and assumed a crouch, the lone pole being carried under the right arm so that when the shoe tips hit light snow and the flakes flew they would go under the arm and along the pole, thus protecting the face and eyes when great speeds were attained. As he succinctly pointed out, there were no goggles in those good old days.

Arthur compared some of the match races run on Eureka Peak to ski jumping tournaments he witnessed in his later years. He recalled that there had been numerous snow benches on the longer match-racing tracks and that he and other riders would run over these at full speed and make leaps of from sixty to eighty feet. Perhaps this speed stuff can be inherited, and that is why two grand nephews, Bobby and Allen Ramsey of Reno, in recent years were listed in the ski racing records of the national and divisional ski associations.

The racing Woodwards' sister, Alice Woodward Meffley, was almost 90 years of age when she passed away.

Alice created quite a stir when national media

quoted her statement that "The Lost Sierra snow-shoers lived in the birthplace of skiing." Norwegian officials were quick to issue statements scuttling her claim, reminding the world that Norway is the true birthplace of skiing. They showered Reno with Norwegian ski posters, with which Alice willingly posed for pictures.

It was in Reno, when modern-day skiing was becoming big business, that Alice Meffley told us about her greatest snow-shoeing thrills.

"Girls who took part in the Johnsville races," she recalled, "wore short skirts, which were strictly a local innovation and very scandalous for those days. We at Johnsville were very good lady snow-shoers. It was great fun. The part I used to enjoy the best was the time when the squads all had been run and everyone at the top of the hill came down in a free-for-all. They used to be hanging all over the hill and, sometimes, it would make work for the doctors."

The Johnsville snow-shoeing fraternity, like its counterpart over in the higher regions, included the salt of the earth. There was Margaret Ellen Moriarty, born in Howland Flat and married to Billy Passetta, a miner who came from Austria's Tyrol, but, nevertheless, knew nothing about sliding down mountains on boards. During February, 1906, what would appear to a present-day slat expert to be an almost impossible feat was performed by Margaret Moriarty. She jumped aboard a pair of traveling snow-shoes and, all by her lonesome, poled across the mountains to Table Rock to visit her sick mother. Not even Snow-shoe Thompson, and he was a pretty good man on a long travel, would have attempted such a feat without thinking twice about it. Then there were two well remembered women: Lillian Langhorst and Margaret Sampell, both school teachers in the snowy days. Lillian became famous for up and marrying Tom Jones a Sierra Valley rancher who raced on buttermilk dope.

Fame also had come to Miss Sampell quicker than the whole town could catch its collective breath. With the citizenry watching, she outraced an avalanche as she sped down from the Four Hills mine.

Then there was 74-year-old James B. Collins, who escaped from Elizabeth, New Jersey, many years ago to find a sanctuary above Johnsville at Nelson Creek. He shakes his head and asserts, "What stories some people do tell."

Alice Woodward Meffley, veteran Johnsville snow-shoe racer, pictured next to Norway ski posters. In 1938 they were rushed to her in Reno by startled Norwegian officials, who, smarting under the Lost Sierrans' claims that they were natives of the "real birthplace of skiing," took immediate action to scuttle those claims and let the public know "Norway has always been the birthplace of skiing." W.B.B. photo.

We can recall the big winter of 1938 when an avalanche took out John's cabin at Nelson Point. Later, when we went up there the following summer, we found him hard at work building a new one, over

30 feet tall. Even James has to admit the snows lie 20 feet up there, because he had built an extra front door 20 feet above the ground level!

References:

Interviews by the author
The Plumas National, Quincy
Count My Blessings, by Fannie Woodward Hunsinger
Woodward Family Genealogy.

Johnsville post office and general store, circa 1925. Lynn Douglas photo.

Tall cabin dwarfs James B.l Collins, whose towering new Nelson Point shelter was designed to protrude above 20-foot snow drifts. Note: Winter door entrance was cut in space atop the 20-foot ladder. WBB photo

13

Threshold On The Yuba

"It was because of the isolation from Downieville, the county seat, that the snow-shoers of La Porte, only some 10 miles away from the Yuba, persuaded the state legislature to take them out of Sierra and put them in Plumas County."

Downieville, county seat of Sierra County, the Gold Rush mecca, stands where two Yuba forks flow together. Downieville Museum 1851-52 drawing.

In all of California there is today no civilization so readily accessible and picturesquely typical of Snowshoe Era mining communities, which bore the brunt of wintry days of '49, than those strung along the upper reaches of the Yuba River's "Grand Canyon." For 140 years this vast canyon has guarded the southern threshold of the Lost Sierra. Here, walled by evergreen forests rising protectively up the mountainsides above the river banks, stands Downieville, the historically appealing seat of Sierra County.

Nearby Sierra City nestles at the foot of the Sierra Buttes, still retaining much of the glamorous atmosphere imbued by the first hardy gold-seeking pioneers. Then, of course, there is Forest City in the Alleghany district, once an annex of the Lost Sierra snow-shoeing empire.

Downieville stands at the confluence of two forks of the Yuba River. Sierra City is perched twelve miles farther up the Yuba's twisting canyon. They both straddle State Route 49, the Yuba Pass highway. Since 1947 the highway has been open to winter travel up through the high elevations to Reno.

Remaining a little world all its own is the Alleghany district, seven miles southwest of Downieville. It can be reached over a windy and round-about mountain road which takes off from the main highway at Goodyear's Bar.

Road builders followed the first miners into the mountain wilderness along the Yuba River, but the going was tough and it was not until 1859 that they completed a passable wagon road connecting Downieville and the Alleghany with Camptonville which, during the early years, had been the stage coach and freighting terminus on the Sierra's western slope for that region. The road was opened to travel in 1859 with a Fourth of July celebration. The first stage that rolled up from Camptonville captivated crowds with its waving flags and banners. It was a great day along the Yuba because the new road meant abandonment of the time-honored pack mule. Those beasts had threaded the narrow trails for nine years. The new service signaled the establishment of faster communication with the outside world. The truth is, the new road did not mean the elimination of snow-shoes in the winter months, but it did assure cheaper delivery of provisions so that, at the very least, there was to be less gouging of those who remained in the high country during the winter time.

The earliest route leading into Downieville was down the steep and narrow Galloway Hill trail, which connected with the Henness Pass route used by the mule-train packers. Henness Pass was a trans-Sierra route used by the emigrants. Improvements were made on it in 1853 through the expenditure of state and county funds. When the siren call of the Washoe silver strike was heard during the winter of 1859-60 it was over this pass that many, first on snow-shoes, and then by pack train, traveled to the new El Dorado, which was to become Nevada's famous Comstock Lode.

Pioneer road builders found slow going as they hacked out wagon roads farther east, up the Yuba, past Sierra City. They eventually cut a route through to the high plateau embracing the lush Sierra farming communities on

Sierra City's Main Street, Highway 49, circa 1900, remained unplowed. Dark building, center background, once housed the Wells Fargo office and U.S. Post Office. Sierra County Historical Society photo.

William M. Stewart grew up in Downieville, went on to become a U.S. Senator from Nevada. Sen. Stewart presented famed mail carrier Snowshoe Thompson's petition for payment to Congress. Nevada Historical Society photo.

Howland Flat, Poker Flat, Newark and Gibsonville. Those mining communities were all famous stops on the snow-shoer's racing circuit. Roads to those towns came up from the natural gateways at Challenge on the west, Quincy on the north and Johnsville on the east. It was because of the isolation from Downieville, the county seat, that the snow-shoers of La Porte, only some 10 miles from the Yuba, persuaded the California legislature, which passed a bill, March 31, 1866, to take them out of Sierra and put them into Plumas County. La Porte thus became the center of Plumas County's "Goodwin Township."

Downieville and the Alleghany district residents were so affronted by the secession of the famous La Porte snow-shoeists, that the miners hanged its legislative representatives in effigy, Tempers soon cooled and today these communities can boast of having been the whilom residence of many civic leaders who thereafter rose to national distinction in various walks of life. The search for gold drew all types of men. They poured into the mining camps. The dominant ones were American youths, trained to respect law and order. They were hard of muscle and self-reliant. Along the Yuba River, and in the surrounding Lost Sierra, it took forceful men to organize the Sierra Nevada's eastern slope.

But the looming mountains to the north provided an insurmountable barrier for connecting links with that original small part of Sierra County. That rough region, centered around La Porte, is now part of Plumas County. It stretches north to Quincy. That remote land is the so-called Lost Sierra, setting for the long-forgotten Snow-shoe Era, land of the legendary snow-shoers.

During the early 1850s, the years that miners had flocked to the Lost Sierra in the search for gold. There had been no direct communication between that fabled gold pocket and Downieville. There were only pack trains or snow-shoe trails climbing to such boom towns as Scales, Port Wine, La Porte, St. Louis,

OFFICE OF THE
Sheriff of Sierra County.
Downieville Nov. 1885
Mr Jake Gould
You are respectfully invited to be present at the official execution of
James O'Neal,
which will take place in the Jail Yard at Downieville, Cal. on Friday, November 27th, 1885, at 2 o'clock. P.M.
S. C. STEWART,
Sheriff of Sierra County.

Necktie Party, Gibsonville Deputy Sheriff Jake Gould, whose descendants still live in La Porte, received this formal, Nov. 27, 1885 invitation to a Downieville hanging.

local governments, frame mining and civil laws to govern themselves and control those of violent inclination.

men who moved mountains

Among such men in Downieville was William M. Stewart. He later became a U.S. Senator from Nevada and presented Snow-shoe Thompson's claim to Congress.

Others leaders we must list included Stephen J. Field, afterwards a justice of the U. S. Supreme Court; Charles N. Felton, who was sent to the U.S. Senate by the California electorate and Col. E. D. Baker, elected to the U.S. Senate by the citizens of Oregon; J. A. Johnson, who left a Downieville law practice to take the California gubernatorial chair; the magnetic William Walker, who later led expeditions to Mexico and Nicaragua and was known as the "Gray-Eyed Man of Destiny." There was a long string of lesser lights.

john mackay

Folklore of the snowy regions credits all these famous men with once having been snow-shoeists. Further, the name of a one-time miner has become integrated into that legend. While the former were all men of education and other attainments before they reached the Yuba diggings, the same cannot be said to have been true of John W. Mackay, who later was to become a world-famous millionaire and a gentleman of many accomplishments. Fortune had not learned to smile on Mackay when he worked the Alleghany ridges, but she must have spotted him snow-shoeing along the Henness Pass one day for he continued on into Nevada and eventually dug millions from the Comstock Lode. He forged on to become world-famous as "Bonanza John."

Mackay had an iron nerve when he arrived in the neighborhood of what was to become Forest City during the summer of 1852. Early accounts reveal the diggings there were struck that year by a company of sailors. Included among them were Bob Ritchie, Little Ned and another sailor named Brown. They were tough customers. Possibly Mackay was among them. It has been recorded that Mackay had arrived in San Francisco several months earlier as a member of a

Mackay statue at the University of Nevada, Reno.

ship's crew. Ten years before that, his parents had brought him to New York from Ireland.

In New York, young Mackay was apprenticed to the steamship building trade. He was a crew member of one of the ships he helped complete. It was thus that the 20-year-old journeyman ship-wright sailed for El Dorado.

At first the new camp reached by Mackay and his comrades was called Brownsville. In the following spring, when the population had grown to some 1000 people, it was renamed Elizaville, as a compliment to a miner's wife. Late in 1853 the name Forest City came into use, but public opinion was split over the best name. There was so much feeling in the matter that a vote was taken in 1854. The majority chose Forest City.

The town had a formidable rival in Alleghany, perched two miles away on the other side of Bald Mountain. During the flush mining years the population peaks fluctuated back and forth between these two competing communities.

Folklore tells the story that the Forest City flat diggings yielded an ounce of gold per day per miner. Drift mining companies were organized and given such descriptive names as Live Yankee, Dutch

143

Company, Empire Hawkeye, Little Rock, Rough and Ready, Can't Get Away, Manhattan, American, Washington, Great Western and Free and Easy.

Though opportunity knocked, no more gold dust than was needed to provide for his immediate wants ever accumulated in Mackay's money belt, but in the Yuba diggings he found good fellowship. It has been said that all during his later years he fondly remembered the time he had spent around Forest City, Alleghany and Downieville. He called it the happiest period of his life. For the Dublin-born Irishman those years were one of hard work during the mining season. He apparently was not one to fritter away time in idleness when the heavy snows closed in. He was living where those along the Yuba encouraged athletic sports and other recreation. They sought to develop the miner's physical resources and to strengthen social and mental faculties during the winter months when for weeks at a time there was enforced idleness. Mackay was never one to waste much time in the grog and gambling shops. Out of the social intercourse of the times in which he participated came a vast storehouse of experience upon which he strengthened his hand when he reached the Comstock.

The Alleghany district boasted of more than famous mines, both past and present. It is a snow-shoeing area that must be seen to be appreciated. In the mid-fifties, the long spruce boards were brought to the region, as they were to the Lost Sierra. The many square miles of snowy open slopes quickened the pulses of the early enthusiasts. Among these was Mackay.

Marveling at the summit of Bald Mountain, which dominates the Alleghany district, we recall the words of Elton O. Carvin, a Sierra County official. He exclaimed that the snowshoer who reached its heights, "must be close to heaven, it is so wonderful."

Looking to the north,, past the threshold of the Yuba, Mackay and his pals could see the snow-covered mountains near Port Wine, La Porte, Howland Flat, Table

Above. Capitol Hotel pictured circa 1900. At the near end of the porch, the sign and stairs leading to the Bonanza Saloon can be identified. Sierra County Historical Society photo.

Many years later, the Capitol Hotel was remodeled to become the Sierra Buttes Inn, pictured in 1985. The porch and upstairs windows remain the same. Day Tschopp photo.

Rock, Gibsonville and Poker Flat. As they turned eastward, stretched before them was Saddle Back Mountain, Fir Cap and Mount Fillmore. Then, beyond to Mount Elwell and the Four Hills country, still further lay Johnsville, just over the ridge. Then marched the Sierra Buttes in all their grandeur. Looking true east the skyline was pierced by Keystone Mountain and English Mountain. Then he could see further on to Mount Pinol, with Cisco and Soda Springs on Donner Pass, just on the other side. If he swung his eyes to the south he could see Emigrant Gap, where the snowsheds of the Southern Pacific were later to be constructed.

Looking true west on a clear day the snow-shoers could see the Sacramento Valley and its mighty river, seen more easily when at flood stage from melting snow. Beyond stood the coast range. Turning northward snow-shoers could see the lonely Sutter Buttes, jutting from the valley floor near Marysville.

Down three sides of Bald Mountain the miners found snow-shoe runs to fill their fancies. Here races were held as early as 1860, but it wasn't until the seventies that events held there became comparable to those staged at La Porte and other mining communities. When these events were announced, all the mines would close down for several days until the races were won or lost. Miners danced every night. Racers came from Downieville, Sierra City, La Porte, Poker Flat, Howland Flat and Johnsville, all bowing to the few top snow-shoers who walked off with the prizes. All knew that plenty of credit went to the dopemakers.

The excitement of match snow-shoe races and news of rich silver strikes in Nevada reached the Alleghany simultaneously in 1860 during the months of January and February. Along with their secret dopes the snow-shoe riders brought verbal reports of the Washoe excitement. It is said that Mackay was more enthused

Snow-shoeing was the only way to attend social functions back in 1908 during deep winter. Outing pictured near Johnsville. Here, fans gathered for races at Snow-shoe Flat.

about the races than about the new El Dorado the snow-shoe riders were talking about. What he had seen of the Lost Sierra from the summit of Bald

Mountain at that time was of greater interest to him than the siren call of the Washoe silver strike. Mackay and a friend, Jack O'Brien, accompanied some of the snow-shoers across the Yuba threshold en route to Poker Flat in order to get a first-hand look at that far-away snow country.

The story of Mackay's trip on snow-shoes to Poker Flat was told the writer by Lenny O'Rourke, always the greatest folklorist of the Lost Sierra. The story had been handed down to O'Rourke by Charles Scott, a pioneer hotel keeper in the classic precincts around Table Rock. By the time Mackay and his friends arrived in the high snow country, the Washoe fever was spreading like a flu epidemic. The more affluent mine operators, fearful of losing their labor supply, sought a cure by merely inveighing against it. They pointed out that, despite the glowing accounts from Washoe, there remained some men who were willing to act the better part by staying at home and prospecting and mining the hills around Poker Flat. To work up some excitement it was then announced that new companies were being organized to seek out new prospects throughout the region and that, "we are not all fools up this way."

The entreaties for loyalty to the diggings and hydraulic workings around Table Rock were unavailing. Mine operators watched glumly while snow-shoes were prepared for trans-Sierra crossings over passes buried in deep snow. Many miners who had caught the fever did not own snow-shoes and their demands taxed the manufacturers of the long boards in all communities.

Left. The Keystone Mill, shown before being demolished by the 1907-8 avalanche.

Bottom Left. When spring came in 1908, the devastation at the Keystone Mill was observed. The ore crushing machine, with its drive wheel, stood fast on its foundation. A safe containing $300 cash and the bodies of several mules were found near the ruins.

Bottom Right. Miners shown breaking trail to the Keystone Mill during the winter of 1906-7

Sierra County Historical Society photos.

"Don't you think it's a great deal better to stay here at home," a mine operator's spokesman said, "than to get on a crazy trot off to Washoe or any other inflated country, and take chances of tornadoes, smallpox and Indians? When a man steps up to one of these hills nearby that has proven rich, and sees all around him plenty more having the same appearance and just as likely to be good, it would seem to be just as well to stay where one is comfortably situated and take the chances, as to go further and more likely fare much worse."

That final plea fell mostly on deaf ears, including those of Mackay and O'Brien, who by then had become infected along with many others. The two pals were not going to wait until the snow-blocked mountain passes were thawed out. They doped up their traveling boards and slid back down the trail to Downieville. Up the Galloway Hill they climbed, then poled along snow-shoe tracks to reach the Henness Pass. Hardened as they were following years of work in the mines, the trans-Sierra crossing was a breeze for Mackay and O'Brien. They reached the snowline just beyond Webber Lake, tossed their snow-shoes into the brush, shouldered their packs and set out on foot through Dog Valley. It was an easy crossing into what then was Utah Territory. Reaching the Truckee River at a point near the present town of Verdi, they proceeded downstream 15 miles, passing the future site of Reno. Then they cut directly south across the sagebrush covered hills to the silver mines.

It is told that as the two pals approached the blossoming mining camp on the slopes of Sun Mountain, they halted at a point where the present Geiger Grade winds around the hills approaching Virginia City. Here they assayed the view and their immediate prospects. O'Brien fished for a four-bit piece, their joint capital, pulling it out of his jeans. Holding the silver coin in the crook of his right forefinger, he adjusted his pack, squared his shoulders and proposed: "What the hell, let's go on in dead broke." With a nod of acquiescence from Mackay, he hurled the coin out and down the mountainside.

Thus it was that the two miners advanced from the golden Lost Sierra Snow-shoe Era into one of silver on the Comstock Lode. To be known as Bonanza John, of Mackay was to become linked, in the American legend of wealth and power, along with those of James Flood, James Fair, William O'Brien, George Hearst, William Sharon, William Ralston, Darius Ogden Mills and a host of other tycoons who swam in the silver flood that poured from the Virginia City mines. Not much of that silver ever stuck to Jack O'Brien. His later years are not a part of this chronicle.

Mackay never did return to the Yuba or the Lost Sierra, but out of that region came the woman who was to become his wife. Her trip was to be made possible by the generosity of the snow-shoers in La Porte, where their progeny was to return 90 years after Mackay had passed that way.

References:

Interviews by the author
Downieville Mountain Messenger
Sierra County Historical Society
Nevada Historical Society

14

The Snow-shoers of Poker Flat

"I gave up telling outsiders about snow-shoes many years ago. It was difficult to explain that they actually were skis. When I told about running 100 feet a second on snow-shoes I could see that people thought I was a mountain story teller from away back." - Will Spencer.

In all the lexicon of the Lost Sierra you can find no peak more famous than that of towering Table Rock, which dominates the heart of the snow-shoe country. According to the Sierra County History, published in 1882 by Farris and Smith, it was down its slopes, above Howland Flat, in 1869 that, "the astonishing feat was performed of running 600 yards in 13 seconds."

Other once-famous mining towns to the north and northwest of the flat-topped mountain include Hepsidam, Whiskey Diggings, Gibsonville, St. Louis and Pine Grove, while on the southern slopes lay Bret Harte's famed Poker Flat.

The "Lost Sierra" is shown in the background of this 1907 photo taken from above the snow-shoers race track at La Porte. Table Rock's distinctive top can be seen on the skyline at right. W.B.B. Collection.

"outcasts", fact or fancy?

Although the Sierra County History does lend credence to the claimed speed records of the snow-shoe racers, the authors of that work sadly consigned Harte's entertaining narrative, "The Outcast of Poker Flat," to what they called "The Liars' Shelf." According to Harte's epic story, written in 1869, when the Table Rock area miners had become possessed with the mania of sliding down hills on strips of wood, the Poker Flat mining camp had once been the scene of a triple hanging. That charge has been indignantly denied by all those who ever lived in those precincts.

Poker Flat may have been on its good behavior when it was sending dopemakers and riders out in quest of honors during the heyday of the snow-shoe tournaments, but it had not always been able to maintain such circumspect characteristics. It's possible that Harte had become familiar with a choice collection of earlier circumstances wherein death was dealt out to some of its more boisterous residents. But there is no record that miners in Poker Flat took to the practice of lynch law.

During the Christmas week of 1857, the snowbound town witnessed an affray on its streets that caused the death of one miner and the eventual legal hanging of another. The ownership of a pair of snow propellors stuck in one of the street's numerous snow drifts is said to have precipitated a fight between Michael Murray and Daniel Sweeney, who went at each other with bowie knives. Cut four times to the quick-flowing red, Murray appeared to be losing the argument when his blade pierced Sweeney's heart.

There apparently being no proof of cold-blooded

villainy on Murray's part, the popular feeling and sympathy were in his favor during the subsequent trial in Downieville. Nevertheless, Justice tipped her scales the other way and he was found guilty in the first degree and an execution date set for January 21, 1859. Three hundred miners, many of whom had snow-shoed down from up around Table Rock, were barred from witnessing the hanging by a National Guard company. Murray declined to address the handful of official witnesses and while prayers were being said by Father Delahunty of the Downieville Catholic church, he was elevated to the next world by a 300-pound weight being let off a trip hook.

Two weeks before Murray's expedited departure from the snowy scene, there had been another knifing in Poker Flat. This fracas, the miners agreed, had been a cold-blooded killing. John Burke and Jimmy Lyons, companions temporarily in the snowbound town, were eating supper in their boarding house when the latter, finishing first, began to smoke his friend's pipe. Burke became offended at this undue bit of familiarity and protested by plunging a butcher knife into Lyons' heart. Justice down in Downieville decided that this was nothing more than second-degree murder and Burke received a short prison sentence.

Haffey/Cosker home, winter of April 1958. Len Haffey, son of Liz Spencer, stands besides the old homestead which is half buried in snow. Richard Jenkins Collection.

Never wilder or more rollicking than its contemporaries, Poker Flat had its beginnings in 1852 and by the following summer there were numerous well-patronized whiskey shops catering to the several hundred miners working the gravel beds in nearby Grizzly Creek. Because of Harte's epic story the area has become irrevocably linked with all folklore involving mining camp lynch law, but to those who were actually outcasts in the Poker Flat winter snows the place will always be more famous for the derring-do of its longboard riders.

Vital statistics of Poker Flat during the winter of 1950 include only the name of Pat O'Kean, a 54-year-old miner, and the availability of some five cabins fit for human occupation. A handful of others did move in to pan for gold during the summer months, but for the winters of 1948-49 and '49-50, O'Kean was the outcast. He lived

Gibsonville brewery, circa 1905, showing owner, William Highland (Will) Spencer, center. Richard Jenkins Collection.

149

there, all by himself, and once each week snow-shoed out to pick up mail in La Porte.

Ninety years ago things were much livelier in Poker Flat and forging ahead to become the camp's most famous resident was Charles Reid Scott. He was a Scottish sailor who had arrived in the mines during 1853. From the very first introduction of Norway skates the following winter, the name of Scott has been linked with the snow-shoe legend and is a continuing factor in its modern-day revival. Charley Scott was among Poker Flat's very earliest combination dope-maker-racers. In fact, he has been credited with having been the camp's first champion.

In 1853 Scott married the town's first lady champion, Ellen Halley, whose father operated a snow-shoer's caravansary known as the Montrose House.

Charley was 39 years of age when he married Ellen and what transpired in the intervening years before his death in Poker Flat in 1900 provides a vivid example of the part snow-shoeing played in the miners' lives.

That Charley was indeed a sturdy pioneer can be seen by anyone who wishes to scout through old copies of the Downieville Mountain Messenger and other contemporary newspaper accounts. His name is to be found among the racing entries up until the eighties.

"C. Scott's dope" was in use until the veteran's passing.

All of the eight Scott children became snow-shoers. The three boys Charles Jr., Walter and Bob became famous racers in their day.

When we talked to Pat O'Kean, Charles was 80, Brother Bob was 67 then. He was one of the more active snow-shoers in the Downieville region. He and a cousin, Jim Dugan, were heroes of an epic trip across the trackless snows of the Lost Sierra.

The Charles Scott family of Table Rock was photographed in this 1906 sequence at Table Rock race course.

Bottom left. "Doping" their snow-shoes for a fast downhill race.

Bottom right. Showing good style as they plummet toward the finish line.

heroic journey to poker flat

In February, 1948, the problem of delivering an emergency message to a snowbound miner in Poker Flat was turned over to the cousins. They were both employees of the U. S. Forest Service and well-versed in the knowledge concerning the trails through the region. Up in the hills was Roy Thompson, then sharing the isolation of Poker Flat with Pat O'Kean. Thompson had to be told that a member of his family down below was seriously ill.

Bob and Jim, both natives of Poker Flat, brought out their moccasins, traveling shoes and dopes. It should be pointed out here that what snow-shoers call moccasins are what skiers call climbers. The moccasins are nothing more than sacks made of bed ticking. They are pulled over the rear end of the snow-shoes and tied to the runner's feet.

Snow was falling in the Yuba Canyon and blowing in grey streams across the ridges above when the two old-timers began their climb at 11 a.m. Nineteen miles distant, with most of the route uphill going, awaited Poker Flat, which greeted the trekkers at 8 p.m., following a trip in the best tradition of the Snow-shoe Era.

The return downhill journey was made the following day in clearing weather during which Dugan recalls, "We had a lot of fun in the new snow because the running was good. When we bent our knees our shoes would turn, just like steering a Cadillac, as easy as that."

That winter of 1948 also saw the Yuba Pass highway opened for the first time to trans-Sierra travel. The author crossed over from his Nevada home to join the district forest supervisor, Frank Delaney, on a ski trip into the snow country at the foot of the Sierra Buttes. Together we traversed the region through to Sardine and Salmon lakes over trails once used by the early snow-shoe packers working the Sierra City and Johnsville snow shuttle. It can be said that the entire region, as well as the Yuba Pass snow country, is one of awe-inspiring grandeur.

a chat with the "best woman snow-shoer"

In 1938 the author had visited Gibsonville to seek out and talk with some real "outcasts" still living in the region. We particularly wanted to talk to Mrs. Elizabeth Spencer, who had been brought to the mountains in 1858. She arrived at the age of two, from her native Hartford, Conn. Elizabeth's parents, the Michael Coskers, during the flush mining years, operated boarding houses in both Howland and Poker Flats. There, in her girlhood, she had participated in the snow-shoe races down the slopes of Table Rock. In 1873 Elizabeth had married a Poker Flat miner named John Haffey. Haffey lost his life in a mining accident 30 years later.

Elizabeth was married a second time in 1906. Out of her wedding trip with husband Will Spencer, the title, "Best Woman Snow-shoer," was bestowed upon the bride. The winter of 1938 had been one of heavy

"Honest" Len Haffey, shown at his wheel of fortune at Virginia City's Delta Saloon. Haffey, who died in September 27, 1971, was 1911 Alturas Snow-shoe Club cross county snow-shoe champion. He was the son of Elizabeth (Aunt Liz) Haffey Spencer, pioneer settler of the "Lost Sierra" during the Snow-shoe Era. Haffey made a sentimental journey back to Gibsonville at 79, accompanied by his daughter and husband, Dee and Bill Bowman and their three children. He must have had a premonition. He died two days later. W.B.B. Collection.

snows through which the Spencers had made a snow-shoe trip to La Porte, some 12 miles by trail from Gibsonville. It was to hear from their own lips the story of that trip and their famous wedding journey that we had come to visit them.

Logs stuck under the eaves braced all four corners of the Spencer home. It seemed scant protection against the snows which that winter had reached a depth of 20 feet. Elizabeth, despite her 82 years, was still the snow-shoeing belle we had always heard about. The blue eyes of Will Spencer, 15 years her junior, sparkled at our surprise.

Their romance had been the talk of the mountains all during the summer of 1906, so that when Elizabeth and Will rode horseback out of Gibsonville late that November, the townspeople suspected they would be married when they got back. Perhaps Johnny Madden, the local dopemaker, had been let in on their secret plans. Anyway, it was he who made two pairs of gaily decorated snow-shoes and

"Aunt Liz" Spencer won the respect of the Lost Sierra miners with her prowess on snow-shoes at an early age. She continued to demonstrate her toughness when she was 80. At that age, in the dead of winter, she hopped onto her longboards and forged into the forest on a solo trip from La Porte to Gibsonville, seven miles over a rugged mountain trail. Widowed early in life by the death of first husband John Haffey, her lively personality soon attracted second husband Will Spencer. Describing her popular devil-may-care spontaneity, Author William (Bill) Talbitzer, longtime Oroville Sacramento Bee correspondent, in his "Echoes of the Gold Rush," under the heading, "Aunt Liz shocked 'Em;" wrote, "The big event of the week, of course, was the dance held at the La Porte hall. Aunt Liz Spencer was usually the belle of the ball. Liz loved to drink the potent liquor of the day and the more she drank the more she danced. She also wore what was called in those days "Bombay Underdrawers"; and she loved to show them to the boys when she got sufficiently into her cups. There was a night when Liz excused herself to answer one of nature's calls and when she returned to the ballroom she had tucked her skirt inside the Bombay Drawers — a sight that all but broke up the dance on the spot." Liz Spencer died in Oroville, July 13, 1942, at 86, and was buried in Gibsonville. Photo, Richard Jenkins collection.

deposited them over the mountains in Johnsville just in case they might come in handy for the couple on the road back.

Elizabeth and Will stabled their horses in Blairsden and then boarded passenger cars of the Northern California-Oregon narrow gauge railroad for Reno. In that Nevada metropolis they were married by Father John Tubman, who gave them an extra blessing for their return trip. Meanwhile heavy snowfall had blockaded the high country to all travel except to those on snow-shoes. The newlyweds were happy to accept Madden's honeymoon runners when they reached Johnsville. Local snow-shoe runners preceded the happy couple and spread the word on all sides of Table Rock that the newly-marrieds were coming. From Poker Flat, Howland Flat, St. Louis, the Bellevue Mine and other points came members of the snow-shoe fraternity to stage a charivari that has become famous in the snow-showers' folklore. The returning couple was met by a large crowd, gathered on the gentle slopes above Gibsonville and, holding hands, they led a gliding parade of snow-shoers down into the heart of that snowbound community. That evening all joined hands to trip the light fantastic toe in the Western House in a manner remindful of the early days when the snow-shoe riders and dopemakers had truly ruled in Gibsonville.

That was the story told by Elizabeth and Will. Proudly they displayed two pairs of beautifully tapered and red-topped traveling snow-shoes. They were the very ones used on their wedding trip of 32 years earlier. The Spencers chuckled about their more recent trip to La Porte. As Elizabeth explained, it had been made because, "We were the only people in Gibsonville all winter. We just had to get out for some dancing after being snowbound for so many weeks."

It was this fondness for dancing and good times that in 1942 was to result, indirectly, in Mrs. Spencer's death. The Spencers had spent the previous winter down below. They had returned for the summer months and were busily anticipating a forthcoming Fourth of July celebration and dance in La Porte, when tragedy struck. Elizabeth wanted her home spic and span for the occasion. While mopping up the floors of her mountain home, the dean of mountain women fell down her kitchen stoop and broke a hip bone. Nevertheless, the doctors who later attended her said that she would have survived had she only been closer to medical attention when the accident occurred. Pneumonia set in while she was being carried out of the mountains. The once-famous Belle of Poker Flat passed away soon after she had arrived at a valley hospital.

William H. (Uncle Bill) Spencer, whose life was intertwined with the facts and legends of Gibsonville in the heart of the Lost Sierra. Uncle Bill, born July 9, 1872 in Gibsonville, lived to be 102 years old. He is pictured on his 91st birthday celebration, Strawberry Valley. It was always his wish to be buried in Gibsonville. When his death came in Oroville in midwinter, 1974, snows blocked the route to the Lost Sierra burial site. His remains were kept in cold storage until the snow melted so his body could be taken to Gibsonville. During Gibsonville's heyday, Spencer was a member of the Golden Anchor Parlor, Native sons of the Golden West. As he reached 100, he recalled, "those were the good old days when a woman showed no more than her ankles and steak was 15 cents a pound."

During our visit to Gibsonville in 1938 the Spencers had laughed heartily about Harte's most famous story. "You can call all us up here who live in these parts, the Poker Flat Outcasts", they chorused. "You don't think for one minute that Bret Harte ever knew anything about this country in the wintertime? We only wish he had, for Harte was a real good story teller."

Spencer told how the early times were always called the "palmy days when dope was king and money plenty."

153

That alone, he believed, should have been sufficient incentive for the romantic writers to have investigated and written about the region. He said he believed he knew the reason why none had ever done so.

"The record shows," he explained, "that the snows of the Sierra were almost impossible to travel over with the web, or Canadian snowshoes. The nearest approach to the Canadian snowshoe ever seen in the Sierra was what we called a `round tramper,' a poor makeshift only used in extreme necessity or by people who were scared of our snow-shoes. Skis did not go by that name, but were simply called snow-shoes and every city man and easterner always associated that name with the Canadian or Indian shoes.

Haffey/Cosker Home- Summer, pictured circa 1890, the Gibsonville residence of John Joseph Haffey and Elizabeth (Liz) Cosker Haffey and family, showing the horse and buggy used by the ladies in those times. Liz Haffey married Will Spencer after her first husband, John Haffey, died. Richard Jenkins Collection.

"I was born and raised right here in Gibsonville, but I gave up telling outsiders about snow-shoes many years ago. It was difficult to explain that they were actually skis. When I told them about running 100 feet per second on snow-shoes I could see that people thought I was a mountain story-teller from away back."

Off course, Will Spencer always told the truth, but the truth sometimes wasn't enough on which the outlander could base a supposition as to his veracity. The reader can draw his own conclusions from the following, carefully recorded random statements by Spencer:

"There was lots of cheering at the races. You don't suppose a bunch of drinking men wouldn't cheer, do you? In the old days no race meet was complete without at least one bar at the foot of the race track. I can remember Bedrock Borrell tending bar at La Porte and how his beer froze. So he poured whiskey in to thaw it out, and so the crowd had its drinks mixed.

"Let me tell you about the famous `moonlight ride' when Andy Hewitt and John Moriarty ran two dead heats and how Hewitt won the final run despite the loss of one shoe as he came down a specially iced track from near the top of Table Rock. It was so cold that night the timers had trouble with their watches, but we always believed they set an all-time speed record.

"The burial of John Kenny was really something over at La Porte in 1890. Kenny was a backer of squads and, naturally, a saloon man. He died during the races. So all the riders stayed in town to take part in his wake. While everyone was getting liquored up in honor of poor dead Kenny, Buck Mullen snow-shoed to Downieville in search of a priest. It took him two days for the return trip because the good father was not an expert on snow-shoes. It was twenty feet down through the snow to Mother Earth, and the miners cut out snow steps leading into that ice-covered mausoleum. Kenny's remains were placed on four pairs of snow-shoes lashed together and drawn to the cemetery.

"The saloon men were benefactors of all boys, giving them three dollars each to snow-shoe to other towns with the word about Saturday night dances.

"The dopemen never forgave Charley Scott for one of his inventions. It was Charley who thought up the idea of perfuming his dope and after that no one was able to smell out a rival's recipe. You know Scott began it all by slipping in some carbolic soap, but others mostly used oil of wintergreen.

"Then there was the dopeman who had the daughter of a mine owner take some sperm candles on her grand tour of Italy to be blessed by the Holy Father. He melted down the candles for his dopes and that was supposed to keep the devil from riding on his riders' heels," Spencer recalled.

Cross-country racing was a specialty of the Gibsonville boys. The town's last champion in those events was Len Haffey, a son of Elizabeth Spencer's first marriage.

Len, in 1911, covered a seven-and-one-half mile course between Gibsonville and Howland Flat in one hour and three minutes.

Later in life he dealt twenty-one in the Virginia City's Delta Saloon, where he claimed his luck was "almost as good as owning a piece in Mackay's Comstock mines."

Len never did participate in the downhill speed events and he had a plausible explanation of why he preferred cross-country running. He said, "Johnny Madden had too many broken-up riders to suit me. His riders often broke legs, arms and shoulders. Those were the final years of racing, and the dopemen were not particular about the condition of the racing hills. There were too many bumps, too many accidents..."

the end of races at table rock

Following 1911 there were no more snow-shoe races in the Table Rock region due to a population decline which began in 1884 when restrictions were placed on hydraulic mining. A fight had developed between the Valley and the Mountain people. Between the years of 1852 and 1881 an estimated 500 million cubic yards of debris washed down the Feather and Yuba rivers into the Sacramento Valley. This debris clogged the Sacramento River and threatened agricultural pursuits. This resulted in laws restraining unlimited hydraulic operations. The mine operators claimed this was "confiscation" of their gold mining rights, but they lost out in the upper courts in 1884. Nevertheless, there was no immediate suspension of mining operations. While the valley residents had been successful in obtaining legislation to "save themselves" from "destruction from hydraulic mining," in every suit against the mining companies it was first necessary to get proper information that the mines were operating and doing damage. This information was very difficult to obtain because the mine owners had armed guards surrounding their mines to ward off any outsiders. Further, the first anti-debris laws required the cease-mining orders to be served personally upon the actual mine owners.

Many mine operators quit the hills for trips to Europe. This temporarily slowed the shutdown of hydraulic operations. But the flush days of gold mining were numbered.

And so were the palmy days of snow-shoes and dopes.

Working after working was abandoned until 1911 only a handful remained in operation. Soon these, too, were forced to quit.

It was the beginning of the end for the Snow-shoe Era.

References:

Downieville Mountain Messenger
Personal interviews by the author:
    Robert Scott, Elizabeth Spencer, Will Spencer, Len Haffey, Pat O' Kean

15

The Ladies

> WORD FROM GRIZZLY — "We have sixteen men at this camp and twelve or more of them are "old bachelors." So if you see any nice young ladies that are on the marry, just tell them to get on their snow-shoes and ride into Grizzly and you bet they will be accommodated.
> —THE DOWNIEVILLE MOUNTAIN MESSENGER, FEB. 22, 1873

heart-stopping lotta

Idolatry of all femininity was proverbial during the Snow-shoe Era in the Lost Sierra. The early goldseekers were chiefly single men, or men who had left their families at home. It was not until late in 1865 that there was an appreciable influx of wives, children and affianced sweethearts into the high elevations of Plumas and Sierra Counties. In the very beginning the softer sex were outnumbered by as much as one hundred to one in some camps. Twenty years later there were many young miners who remained firmly convinced that anything resembling an even balance between the sexes never would be struck.

Throughout the late sixties and seventies the snow-shoers' sweetheart was Lotta Crabtree, whose lithographed likeness decorated many an isolated mountain cabin. Lotta, a gay little shoot with a most sweet blossom, had engrafted herself onto the miners'

Lotta Crabtree, the snow-shoers' sweetheart, whose picture decorated many lonely miners' cabins. W.B.B. Collection.

hearts during the mid-fifties. With the solid apple-bloom of old Lancaster in the cheeks and possessed of some added saucy merriment, she was seven years of age in 1854 when she arrived in Rabbit Creek with her mother, Mary Ann, to join her father, John Crabtree.

Due to a water scarcity, that mining season witnessed depressed financial times and many, the Crabtrees included, bemoaned the meager pickings in their long toms and sluice boxes. Particularly were the Crabtrees forced to prune down their mode of life so that when the snows came and all trails were closed it became necessary for the mother to maintain the family by opening up a boarding house. Dancing, amateur theatricals and snow-shoeing were fast becoming inextricably interwoven into community life. Out of that combination, little Lotta was to become the most popular and famous soubrette of the nineteenth century.

Lotta soon had won the hearts of all the hearty eaters and love-hungry patrons of her mother's snowbound boarding house. She was a dancing fairy, with a repertoire of songs and mimicry, whose fame soon was carried down below by the early snow-shoe expressmen. It was at this time that Lola Montez, the wonderful spider dancer and somewhat famous erotic genius, was creating such a profound sensation in the susceptible breasts of the Downieville miners that her name was immortalized in the christening of Mount Lola and also two lakes near Soda Springs. Quickly as genius ever is to detect its kindred spirit, Madam Lola set out for Rabbit Creek in the spring of 1856 to engraft herself onto the Crabtrees. There she was strangely touched by the elfin ways of little Lotta, who by that time was considered an accomplished actress. Lola was determined to take the youngster with her when she rode out. Lola would dance and sing for her little new-found friend. She pleaded with the parents to let her take Lotta on an Australian tour. She wanted to train the little blossom for a bright career.

But Mrs. Crabtree had an Italian reason for refusing. He was Signor Bona, who went by the name of Mart Taylor and operated a saloon and a little log theater, where music was abundant and silver coins and gold nuggets sometimes showered down. It was on the occasion of a rivalry between Signor Bona and a visiting theatrical troupe that little Lotta gained her first newspaper notices. The touring thespians, featuring a child prodigy, had declined to rent the theater at Bona's price. The troupe, seeking to avoid the demanded Italian tribute, had engaged a hall across the way. Signor Bona, himself a violin soloist and experienced entertainment caterer, sent over to the Crabtrees for the very special loan of the apple of their eye, Lotta. His request was granted.

To act as wardrobe mistress the services of Mrs. George Babb, a New England lady who was well-known for liking the better things of life, were then obtained. While her nimble fingers cut out and

Famed Soubrette - Lotta Crabtree when age 7, arrived in Rabbit Creek on horseback. She was brought by her mother, Mary Ann, to join her father, John Crabtree, a miner. Above, as she appeared in San Francisco. W.B.B. Collection.

stitched a green velvet costume, Signor Bona scurried about to locate a shillelagh and other properties in order to provide a suitable background for his little star's repertoire of Irish reels and songs.

Of Lotta's debut an old newspaper account, using her press release, said:

> "Thirty dollars in silver rang upon the stage from the gruff and grateful throng of miners, who, utterly ignoring the performance across the way, crowded to the debut of the Italian's protege. The next morning the Rabbit Valley Messenger went singing out over the valleys of the Golden State carrying an echo destined to grow into a national note."

Of course, the Crabtrees were not going to divest themselves of any such prize, not even to Lola Montez. Soon Mrs. Crabtree and Signor Bona set out on muleback with their little charge, to answer the peremptory calls of other isolated camps, where the miners would have the best thing for their money. They sought a "fresh taste of nature in the charming little new sensation, lithe and laughing Lotta."

The little group's trail led them through St. Louis, Howland Flat, Gibsonville, Onion Valley, Saw Pit Flat and down to the Feather River. Little Lotta entertained in barrooms and what in those days passed for theaters along the way. All the while Lotta was storing up experience for a career which carried her to international fame and a $4 million fortune.

In later years, Mrs. Babb, who had accompanied Lotta and her mother to St. Louis and Howland Flat, was to recall that when the miners showered down silver and gold acknowledgements it had been a "thrill for the child to stop singing and dancing long enough to stoop down and pick up the money." Further, it was Mrs. Babb who often told how Lotta used to entertain at

Lola Montez
A Snow-shoe Era miners' favorite, memorialized by the naming of Mt. Lola, near Truckee, Calif., Lola Montez was the famed "spider dancer." She was also known as the Comtesse de Landsfeld, a title bestowed by King Ludwig of Bavaria. In 1856 while she was living in Grass Valley, she heard about Lotta Crabtree and immediately rode to La Porte, where she spent many days singing and dancing with the young enchantress. She made a vain attempt to take the "snow-shoers" sweetheart on her Australian tour, but Lotta's mother, Mary Ann Crabtree, would not let Lotta out of her sight. A couple of years earlier, July 13, 1854, Lola had made the headlines in the Grass Valley Telegraph, when she set out on a horseback trip across the Sierra Nevada to take a look at the Truckee Meadows, where Reno stands today. James Delavan, MD, who accompanied the bloomer-clad entertainer, wrote an account for the Adrian, Mich. Expositor, telling of the hardship she endured on the trip. He described her visit to the celebrated Donner Cabin, scene of epic starvation and death eight years earlier. He wrote, "The eccentric Countess brought away with her some of the human bones, which are quite plenty there." The trip took several several days on horseback each way. They arrived back in Grass Valley, July 28. The arduous ride gave ample testimony to the frontier spirit of the famed dancer.

the "snow-shoers doings." There are many other legends linking the actress' life with the Lost Sierra.

Before their passing we talked to old-timers who claimed to have played with Lotta as a child. They often looked at Table Rock and wondered out loud why it was not graced with an observatory. They all seemed to have a firm conviction that among Lotta's bequests, and she did leave the $4 million for distribution, had been one providing for the erection of just such a remembrance of her days in the snows of Plumas and Sierra.

Naturally, there were many other maidens who became entwined in the snow-shoers' hearts. We would like to believe that some were enough "on the marry" to have snow-shoed into the Grizzly Mine and have been accommodated by the isolated bachelors, but there is no record of such a rush having taken place, although numerous other stories are told of women who ran down their men on snow-shoes.

women held their own!

Only once during the Snow-shoe Era did any miner have the temerity to question the propriety of feminine participation in snow-shoeing and he was quickly knocked down by the Mountain Messenger, which tartly stated: "We don't agree with our correspondent that ladies are out of place on snow-shoes. Properly dressed, we can see no reason why they should not participate in what are termed manly sports, as well as their fathers and brothers."

The question arose when a correspondent from Howland Flat commented upon the snow-shoeing debut of Miss Lottie Joy, who a few weeks previously had won the first championship belt of the Alturas Snow-shoe Club.

"Our feminine performer on snow-shoes," the correspondent had penned, "made her debut not as a snow-shoeist, but as an accomplished gymnast. Her gyrations were truly sublime. How I wished for a Raphael just at that time, to paint in colors undying the beautiful houri trying to keep on the track as well as her shoes; but, would you believe it, the fates would refuse. But I will endeavor to give my opinion on the subject without getting poetical. I think such athletic sports inappropriate and out of place for the gentler sex."

No sooner had the newspaper's subscribers digested both sides of the argument than a feminine correspondent, signing herself "Mischievous" but firmly believed to have been none other than Miss Joy herself, expressed the ladies' viewpoint of the sterner sex on snow-shoes. There was nary a male in the mountains who ever had an answer for what she had to say. Her somewhat remarkable bit of correspondence follows:

"I have been the victim of all kinds of changes that pertain to this uncertain climate, our unchangeable habitations at this season leaving us at the mercy of every infliction that winter may see fit to impose. My spirits and the mercury have fallen together, below zero for the past three months. Without the renown which attaches to a visit to the North Pole, we have enjoyed a luxuriance of the beautiful snow, and been deprived of solar light and heat for months together, living on salt rations, last year's newspapers, old almanacs and musty books. So long submerged in the snow, I wonder if we have no adaptabilty to become amphibious, no inclination to the torpidity of the bear, or the contented stupidity of the Esquimaux.

"During the prevalence of these dreadful times, and to add to the misery of indoor life, a peculiar malady or mania breaks out in which the whole community, being wholly confined to the sterner sex, and which, like the itch, occupies the whole attention of those afflicted. Though not positively dangerous, its symptoms are often alarming. A diagnosis might not be uninteresting to the public and the profession. Even physicians themselves are occasionally attacked with it, and one editor, who strayed here a short time since, fell an early victim, and for a time was given up as lost. The first symptoms are those which affect the brain. The patient generally seeks a druggist and invests sundry small sums in beeswax, rosin, tar, asphaltum, spermaceti, gum camphor, barberry, shellac, gum arabic, gelatin, etcetera, all of which, duly numbered and made into a kind of wax is vulgarly called `dope' which he rubs over the soles of his snow-shoes, and immediately thereafter tries to ascend Table Rock.

"The first trip he generally gets about halfway up, and looking down upon the pygmies crawling below, draws himself out at full length, mounts the untamed steeds and the next instant disappears beneath the snow. But he comes out again all right, and after several contortions, such as would immortalize a dancing dervish he reaches the Hibernia ground.

"A little more doping before the second ascent, and a little further up he drives down the cap on his head and turns around. He mounts. The next instant you see him standing on his head in a pair of snow-shoes about 12 feet long, the bottoms of which flash an instant in the sun. The next you see is a chap about his size standing a mile below, down by the Hibernia, as before, putting on dope `No. 3.' The symptoms are all the same. Sometimes there are a dozen crawling up and others shooting down the hill, all crazy, funny fellows. Every one of the fastest with a piece of dope, which he tries to keep the others from smelling, and all the others trying to get a smell from him.

"The dream of the alchemists who have died without discovering the true afflatus by which the diamond is crystallized, has only given place to a new inspiration after hidden science, which is now being concentrated upon the fastest wax and Prof. Ayer, Victor Carrington and the nitro-glycerine man have each his peer at Howland Flat, who will, no doubt, some day become known to science and fame. Perpetual motion could at once be established by our savants, if whereon to rest their lever could be found, or, in other terms, if a track long enough were only laid out. The sunlight travels round the earth at a tolerably rapid rate, but were there only a snow-shoe track, not only would the fast dope keep up, but outrun it and give time besides for making calls on our benighted friends at 'Greenland's icy mountains and India's coral strand.'

"An outsider might naturally suppose that all this ado is about nothing; that the snow-shoes and diamond business are both played out, won't pay. No such thing. Those who have the fastest dope are the richest men in the country. I have been on the track and heard them bet thousands and millions of dollars, though they rarely put up the money, owing, I suppose, to the inconvenience of doing business at this distance from their deposits. But they win a great deal of money besides the regular purses run for, with lots of gold and silver watches thrown in. The champion, who is generally a bow-legged fellow, with a parboiled nose, gets presented with a belt made of leather and studded with variegated stars, such as are supposed to have been seen nowhere else except in the snow-shoe constellation. This belt is drawn so tightly around his waist that both his eyes protrude like big acorns, and can be seen at a great distance when he blows his nosegay coming down the track.

"The free balls, for the benefit of our visitors, are models in their way, the demolishers of caste, and the winners of the day are all petted by the ladies without reserve, especially when Leap Year permits this general license. Still, I regret having to remark that the time of many of the gentlemen present was about equally divided between ourselves and the inevitable bottle, the great social curse and associate of every other foul vice.

"We have survived the weather and the races. The latter have, to a considerable extent, offset the former, and given us ladies a chance to sun ourselves between storms, for all of which we are duly grateful."

At La Porte in 1867 the keen eyes of Quicksilver Charley Hendel had observed the fair sex sliding about and his pen flowed with praise.

"The ladies' racing," he wrote, "was the most exciting and interesting of the week. All over the mountain could be seen women practicing. In one direction a little innocent miss, with her scarlet dress fluttering in the breeze; in another a blushing maid cautiously gliding down some gentle slope; and still in another some harum-scarum, don't-care-a-cent woman would dash down some steep declivity and by losing equilibrium come to grass, or rather to snow, with far more regard to ease, than elegance."

That had been an unprecedented winter. Soon after the weather had begun to moderate sufficiently to tell on that "little dab of snow" surrounding Table Rock. It was Quicksilver Charley who thought up the idea of a dinner party to honor Lottie, the lady champion.

The town of Grass Flat was beginning to look cheery, what with Corbett's Hotel beginning to be exposed to view. In the hotel's second story a table was set. Just as the party was sitting down, one-half of the flooring gave way, precipitating many of Lottie's admirers and some specially prepared chicken fixin's to the dance hall floor below. Among those who crashed down was a miner named Tom Fennimore, who claimed he had made the descent with a coffee pot holding five gallons of boiling brew without spilling a drop. One of the snow-shoeing belle's more ardent swains, Fennimore was also notorious for his addiction to stretching the truth and he embellished the coffee pot story with "it's the truth, by God, and if my brother was here he would prove it." However, while fearless Lottie had simultaneously leaped out the window to safety in the snow below and several guests were seriously injured, a contemporary editor evinced more concern for the edibles destroyed and $100 worth of crockery smashed to smithereens.

courtships in the snow

Putting snow-shoe tournaments on an organized basis brought rapid changes to mountain social life. By the winter of 1870 the mining camps' original boy and girl crop was reaching the courtship age. Too, steel had linked steel on the transcontinental railway route and there were many newcomers, including a sprinkling of the fairer sex, who, thank heavens, were for the most part definitely "on the marry". Doting parents of teenagers were encouraging young romance with card parties and other simple amusements interspersed between the excitement of the snow-shoe tournaments. At the same time the fair newcomers stirred the manly veterans of lengthy bachelorhoods. It was becoming fashionable to woo during the long winter months and then to marry and honeymoon in the snows of spring.

Among the many passengers carried to California in November of 1869 by the new Central Pacific Railroad was Mary Ann Lewis, an attractive 17-year-old lassie who had come all the way from her native Wales to visit a brother in the Poker Flat mining region. While such a journey was an adventure in itself in those days, the trip was as nothing compared to what Mary Ann was to experience after she quit the iron horse and steel rails in the Sacramento Valley.

Brotherly letters had told her about the beauties of the Far West, but there had been little mention of the snowbound isolation of the High Sierra. As she was to recall in later years, and her passing did not come until 1949, the little Welsh beauty arrived in Marysville firmly convinced that a short stage coach journey would place her in some kind of sunshine-drenched El Dorado, a combination place where oranges grew, gold was picked up, and a girl could find matrimonial opportunity.

So it was late November 1869 that Mary Ann purchased a ticket and set out by stagecoach for La Porte.

The road was rough, then became steep as well as the stage reached the western slope of the Sierra. The driver, a guard and several Chinese, naturally rode on top. Mary Ann was packed inside with several miners. She found it almost impossible to avoid being squeezed at times. The squeezing was the result of the coach's swaying rather than any extracurricular miners' prospecting.

From Marysville through the lowlands it was an interesting drive, but there was snow on the ground when the mountains were reached and Mary Ann opened her eyes. After passing through Strawberry Valley, the passengers on occasion were required to step out and push the coach up numerous slippery grades. At other times the curves were so sharp that all Mary Ann could see of the six horses were the last two horses' tails. The 68-mile journey took 18 hours. Mary Ann was a weary and shaken up little girl long before it ended.

As usual the whole town of La Porte came running out to meet the coach and crowded around like children at a Christmas tree. All were curious to see what it had to offer. The Downieville Messenger, correspondent pad and pencil in hand, pushed his way around and jotted down everything he picked up. The hotel owners were out in force and all anxious to accommodate the passengers. As for the single men, it can be said they were wife prospecting.

There were also other possibilities in such mass assemblies. The miners' bachelorhood always remembered what had happened on another occasion when a beautiful young lady was stepping out of the coach and her hoop skirt had become entangled with the boots of a male passenger. When both her feet hit the ground, her hoop skirt was still caught in the coach. It had been a cruel disaster, and many miners smiled in anticipation, but there was no re-enactment when Mary Ann stepped down.

Among the observant bachelors was Joel Bean, a 34-year-old former native of Carratunk, Maine. He had come to the mines in 1856 and now was showing great promise as a mine superintendent, He watched in silence as Mary Ann leaped into her brother's arms. He knew the brother lived in St. Louis, where for Mary Ann the snow was something new. Soon he was among the many miners who were snow-shoeing to her brother's home and sitting around in open admiration.

In the snow-shoeing matrimonial sweepstakes it was Joel Bean who had the greatest derring-do. He taught the pretty Welsh miss how to snow-shoe and revealed to her the mysteriousness of dope. He ran a fast race and won his beautiful prize at the winning poles of matrimony the following March.

The nuptials had been celebrated following the annual tournament of the Table Rock Snow-shoe Club and it was only natural that many members of the fast dope fraternity were on hand to wish the happy couple good luck as they poled away to establish a home in Scales, 15 miles distant. That was where Bean held sway as superintendent of the Cleveland Gold Mining Company. The mine was so named because it had been financed by Ohio capitalists.

The little Welsh bride proved to be something of an extra added asset to the social life of the mining community. The newlyweds established their home in a company-owned house, where six children were delivered to bless the union. The family continued to live in that snug home until forced out into a blustery winter night by fire in January of 1898. A new home was constructed and the Beans continued on until 1911 when mining operations petered out.

The family then moved to Strawberry Valley. There the woman who had come to the snow country as a young girl "on the marry" became the recognized dowager of the snow-shoers. She rounded out her life span there at 97 years of age.

Hers was a long, full and productive life, one that can be documented as typical of so many other mothers who built the American way of life on the Western frontier. All of the six children she bore came into this world in a rugged mining camp, minus the assistance of a doctor. There never was a medical man in that camp when Mrs. Bean needed one. It was always her proud boast that "a mid-wife was always good enough for me."

On three occasions the Bean home was snowbound. Once the stork beat the mid-wife who was racing on snow-shoes to the scene. The first-born, Abraham Lincoln Bean, arrived the evening of Dec. 23, 1771, but the mid-wife had been forewarned and arrived through a snowstorm in plenty of time.

However, things did not take place as anticipated for the arrival of Vernon J. Bean, who always was know as "Santa Claus' Surprise," because he arrived unexpectedly Christmas Eve, December, 24, 1890, while his mother was decorating a Yule tree. That was the time the stork outglided a mid-wife despite the slick coat of lightning dope on her snow-shoes. The other snow baby was John A. Bean, who came with the season's first snowstorm on Nov. 26, 1876. He became a well-known snow-shoe racer for the Alturas Snow-shoe Club. The other Bean youngsters, all experts too, were Morgan, Amanda and Laura.

Now just another ghost mining camp, Scales in its heyday, stood at the conflux of the main snow-shoe and mule-packing trails. They led up into the Lost Sierra from Downieville, 18 miles to the southwest. Northward from Scales it was another four miles to Poverty Hills, then four to Port Wine and six more on into La Porte. During the mid-eighties the camp became a natural stopover for all snow-shoers headed to and fro between La Porte and Forest City, perched in the Alleghany region, where racing tournaments had become the big winter social activity following the success of quartz mining activities in the snow country south of the Yuba River.

Because her husband was among the more prominent mine superintendents, Mrs. Bean often was an overnight hostess to many snow-shoe travelers. Guests were wont to display their secret dopes on the way down. On the return trip they would bring copies of the Sierra Tribune, published in Forest City. The newspaper contained race results and other snow-shoeing stories. Leap Year, 1884, was enlivened by masquerade balls and other high jinks in all the snow-bound mining camps. These social activities were well and colorfully publicized in the mountain press. No editor liked to be outdone in snow-shoe reporting, especially the publisher of the Sierra Tribune, who on Valentine's Day that February 14, hand-pegged a love-in-the-snow classic for his waiting press. The account, a copy of which was found among Mrs. Bean's most treasured memorabilia, reads as follows:

"One evening last week a small party of young folks were returning home from an evening's visit and while descending the Bald Mountain track, a certain fellow endeavored to, we will speak plainly, kiss his girl. After expostulating on the subject for some time, and finding his entreaties in vain, he finally concluded to have his kiss anyhow. The young lady was firm, however, and stood her ground. A fight took place and at the beginning of the first round our hero and heroine both landed several feet below the track in a soft bed of snow and being unable to get out were obliged to call for help, which soon came, much to the chagrin of the two adventurers."

What would today's younger set think of having to travel 10 or 12 miles across the snows to attend a dance or card party? Yet that is just what sparking couples as well as matrons did frequently in the early days. Jolly good times they had too.

There are stories of the miners' adulation of the snow-shoeing ladies beginning as early as 1856 when a Grand Christmas Ball was held at Rabbit Creek. The dance, for which John "Buckskin" Porter, the pioneer snow-shoe manufacturer, was among the floor managers, attracted just about every miner and lady from miles around. Of course the fair sex was outnumbered ten to one and it took cool heads and strong arms to keep the exuberant miners in line. Of that evening it has been recalled that "there was a Christmas shining light and nobody thought about sorrow and sin. No smooth-faced gents with soft front shirts were at the dance that night, and many drank too much."

Christmas, New Year's, Washington's Birthday and St. Patrick's Day always were the big social dates on the snow-shoers' calendar. As each season progressed the snow usually became deeper with each passing week until just about everything in sight had been covered in white by the time St. Patrick's Day rolled around. For instance in 1859 there must have been at least 18 feet of snow on the level for the Irish celebration in La Porte. The Mountain Messenger reported special arrangements had been made for the convenience of visitors: "A tunnel has been run along the sidewalk from the Union Hotel to Conly's Banking House; another one from the Union to the El Dorado Saloon, and a third from Evans and Morley's butcher shop to Freeman's Restaurant, all of them at least 10 feet below the snow surface."

The Irish parked their snow propellors above and came down below with shillelaghs and tripped the light fantastic to the "Wearin' o' the Green."

Plenty of advance notices were given to all social doings in the Downieville Mountain Messenger. In mid-February of 1863 the paper reported that the "heaviest storm of the season had set in" but was not inconveniencing the snow travelers planning trips to Pine Grove, where "Henry Daugherty, proprietor of the Table Rock Hotel, a clever gentleman and enterprising landlord, promises grand dancing entertainment at Union Hall on the occasion of Washington's Birthday."

The winter months were ones of snow-shoe racing and downhill tournaments, dances and social intercourse with the participants traveling across the snows. It was in such an environment that Mrs. Bean became an acknowledged leader during her many years in Scales. Her influence can be noted in a

Downieville Mountain Messenger story of 1873, telling how at "Scales Diggings the boys are enjoying themselves as best they can while waiting for water, indulging in snow-shoe racing, singing schools and dancing schools."

By the time Mrs. Bean reached her twenty-first birthday she had acquired the numerous skills necessary to be an expert mountain housewife. She passed along her new-found knowledge to other young women who by that time were setting up housekeeping in ever-increasing numbers. True, the population of some camps was declining, but there were still many men who hungered for the association of good women. Further, there was culmination of many romances among the native-born younger generation.

More experienced matrons taught the brides what they must know to satisfy their menfolk. There were berries to be canned, vegetables stored in root cellars, bacon and hams prepared, beef to be corned and numerous other chores learned in order to be ready for the long winter months. Nursing skills also were among the essential skills. Sickness often stalked among the miners' children during the snow season. It was then that the married women won praise for much good work.

The mountain women were and always will be proud of their contemporaries' part in the conquest of the Lost Sierra. Among these is Marguerite McMahon Delahunty of Whiskey Diggings. In 1900 she married Frank Delahunty following a snow-shoeing romance begun six years earlier. Both were born in the high country. Marguerite is still affectionately known as the "Balsam Girl" because of her knowledge of the fast wax art. Delahunty, until his death, was a mining engineer of some repute.

Children went to school, visited friends and attended social gatherings on snow-shoes. Above three tots, from left, Verda Smith, Elmer Schubert and Sylvia Robinson display their Snow-shoe Era outdoor dress. The 1903 photo was taken at a Gibsonville Ridge, between La Porte and Gibsonville in the Lost Sierra.

Insert. Even the very young were skilled at using snow-shoes.

As a young socialite of the snows during the gay nineties, Mrs. Delahunty traveled about on snow-shoes to the racing meets and mountain dances. Always pursuing her were snow-shoeing beaux. Delahunty ran her down and wrung a "yes" from her in Secret Diggings. The couple established a home in the very heart of the Lost Sierra and their union was blessed with several offspring.

"It was the kind of life," she asserts, "that brought out the best in any woman.

"There were children to be raised in the American way. Girls to be schooled in the domestic arts and deportment, and growing boys taught the evils of life in the saloons. The wonder of it all is that so many youngsters turned out as well as they did. Naturally, many of the young men did mingle in the drinking and gambling places. After all those places provided the principal recreation for the menfolk. But we women folk, including our daughters, provided more stability in the form of simple amusement with card parties and musicals in our homes.

"Of course, all the girls could snow-shoe. There were few men who ever escaped because they were on the marry, too."

References:

The Forest City Sierra Tribune
The Downieville Mountain Messenger
Author's interview with Mrs. Joseph (Mary Ann) Bean
Author's interview with Marguerite McMahon Delahunty

Ladies, as shown in this 1908 photo taken at La Porte, were well accustomed to longboards during the Snow-shoe Era in the Lost Sierra. Snow-shoes were used for traveling to neighboring towns, shopping, church attendance and community social events.

16

Miners Unite in Spirit

Miners, merchants, editors, preachers, saloonkeepers and all other professional men joined in "Hail To The Clampers". "The rollicking fraternal group appeared in 1857 and became the best loved group in the California mining camps.

Members of the original snow-shoeing fraternity were all strong, self-reliant and dauntless men. They had to be intrepid to build the way of life they did in the snows of Plumas and Sierra.

The first true social stability in the mining camps stemmed from the presence of members of fraternal organizations such as E. Clampus Vitus and itinerant ministers of the Gospel. Early day life was unquestionably lusty and raw, but there were those among the first comers who believed in fraternity and were God-fearing as well. To offset the prevailing vices of gambling and drinking in the mining camps there was membership in fraternal organizations such as the Masonic and Odd Fellows' lodges and verbal encouragement from both Protestant ministers and Catholic priests.

The best minds of the mining, business, fraternal, press, political and church world of the Snow-shoe Era were heavily drawn upon in order to bring order out of chaos during the Lost Sierra's early and lusty years. Evidence this was true was emphasized with the region's response to the death in 1897 of Benjamin Wilcox Barnes, of La Porte. Barnes, a native of New York, had been among the organizers and first officers of the Alturas Snow-shoe Club. At the time of his passing he was about 76 years of age. He had come to Rabbit Creek in 1854. His career as a private citizen and a public man was well known. First he was interested in mining activities thereabouts and when the camp's name was changed to La Porte, it was Barnes who constructed a pipeline to supply the growing community with water. He took much interest in all matters of public concern. Whatever affected the prosperity of the mining camps seized his close attention. He was a member of the Jefferson Masonic Lodge, organized, May 8, 1856; a worshipper in the Methodist Episcopal Church, dedicated during the summer of 1858, and a charter member of the Alturas Odd Fellows' Lodge, instituted in December of that year.

A Republican of the old school, Barnes became a prominent figure in county, district and state conventions of his party. In 1871, having received a plurality from his snow-shoeing constituents, he was elected a member of the California Assembly, and as such was said to have served with ability and distinction. On questions of the day, the snow sports included, he was an able and frequent contributor to the local press. Among the snow-shoeing fraternity his name had occupied a high and honorable station.

Barnes' passed away April 7, 1897, in his snowbound La Porte residence. The news, though painful to the snow-shoeing fraternity, was not a great surprise, for Barnes had been known to be in failing health. All that winter the mortality among the old pioneers had been high, but recurring storms had prevented many from paying last respects to old friends. Now there were those in attendance at the death watch who were determined that such should not be the case for their old friend.

An immediate decision was reached by Barnes' widow. The departed man's fraternal brothers ordered Masonic rites to be held the following day. Three hours later there were snow-shoe messengers poling throughout the mountains with the sad news. By daybreak the following morning there were many snow-shoers trekking toward La Porte, while down in Quincy, 35 miles away, the Plumas National reported:

"The funeral takes place at La Porte this afternoon and, notwithstanding the depth of snow, will be largely attended by the people of that part of Plumas and Sierra counties, who have known the esteemed and along the route toward Nelson Point who had other ideas about the difficulty of crossing the dividing mountain ridges. Too, there were some likewise minded down in Downieville on the southern threshold. Before nightfall, April 7, there were small parties snow-shoeing toward La Porte from both these communities. Following overnight stops en route their ranks were bolstered by many others from the numerous mining camps in the high country. By funeral time there was a forest of snow-shoes up-ended in the snow banks of La Porte.

Meantime preparations for full fraternal rites had

Rev. George M. Darley, D.D., left, best known for his preaching to snow-shoers in Colorado, followed by an associate, took his ministry into the mountains.

deceased nearly 40 years. It is to be regretted that the banks of snow between here and La Porte render it impossible for his old friends in this part of the county to attend the funeral of their pioneer friend."

Nevertheless, there were snow-shoeists in Quincy been carried out by the Masonic and Odd Fellows' lodges. In the cemetery on Waite Hill the lodge members had dug through 12 feet of snow to reach Mother Earth and prepare a grave. The remains of Barnes were lashed to a sleigh, which was drawn by snow-shoe-equipped horses to Fraternal Hall on the banks of Rabbit Creek. Following fraternal services the cortege, with Masonic members dressed in full regalia, moved across the snows to place Barnes beneath a mantle of white.

The task of framing suitable memorial resolutions was given to three contemporary Alturas snow-shoers: Quicksilver Charley Hendel, the sport's eloquent chronicler; John Hillman, a dopemaking dean, and James Jones, an early days racer. Their combined effort; a true thought from the snow-shoers' hearts, follows:

"Resolved, that we recognize no higher compliment to living, no prouder eulogy for the dead than this, in him we ever found a true, earnest and faithful Odd Fellow, and these things we say now, of our departed brother in all fervency and devotion, that he was well worthy of them. Resolved, that our hearts are filled with profound regret by this deep affection, and we mourn the loss ourself has suffered. We forget not his family, weary, stricken and distressed; we console them in their bereavement and sympathize in their sorrow."

Snow-shoer friends of B.W. Barnes, on hearing of his death, trekked from miles around to attend his funeral, despite the deep snow piled up around La Porte, April 8, 1897. Above, left, they marched in a funeral procession. At right, formal Masonic rites were performed in front of the Methodist church. The grave had to be dug down underneath 12 feet of snow.

"Whereas, it has pleased the Almighty Ruler of the Universe to remove from our midst a worthy charter member, our well-beloved brother, B. W. Barnes, therefore, be it resolved that by the sorrowful event, society has lost a good, worthy and useful member, the state at large a true and noble citizen, the unfortunate and distressed a friend and benefactor, while his family have been deprived of a worthy husband, and the lodge a member well worthy of all confidence and esteem.

The Masonic fraternity reached the foothill country in 1851 when Mountain Shade Lodge was organized under dispensation. The following summer the Rev. R. R. Dunlap was preaching Methodism to the religiously inclined, while simultaneously a Catholic church building was erected with Father Thomas Dalton as priest. Downieville's next stabilizing influence came when Sierra Lodge of Odd Fellows was organized in 1854. It was followed two years later by the Blue Range Encampment. Odd Fellows and Methodism came simultaneously to the Alleghany District in 1854 when the Forest City Lodge was instituted and a preacher set up a permanent pulpit in that town.

It was Methodism which first held the field in the high elevations between the Yuba and Feather River canyons. First there were itinerant preachers. Then came a permanent organization at Nelson Point in 1854. The Rev. J. C. Gentry, a studious, pious disciple of the Methodist Episcopal faith was the Gospel's first dispenser. Then in 1858 came the Rev. Philetus Grove, who traveled through all the camps up around Pilot Peak and Table Rock and whose faithful labors lived on to bless for many years all the dwellers in the snowy regions. Further, there were traveling priests of the Catholic faith. The Men of God were aided by the citizens with true California liberality. Churches and missions were built where possible.

Early ministers and priests were not necessarily expert snow-shoers. Nevertheless, they did travel across the snows to visit and preach in isolated parishes. There are numerous legends and old newspaper accounts rich with such anecdotes.

While the crowds were in La Porte for the races in March, 1890, the death of Jack Kenny, a snow-shoe squad backer, was almost the end as well for a Father McClare of Downieville. The racers stayed in town to bury Kenny in proper style. The wake and services were a big event. They were somewhat delayed while Buck Mullen, a local snow-shoe messenger, traveled down below to obtain the priest. Father McClare was not a young man. It took him two days to climb the hill. Mullen, seeing the priest needed a stimulant, bouyed up his charge with liberal potions of mountain whiskey. Meanwhile the snow-shoe racers had deposited Kenny's body in a convenient snowdrift and then shoveled down through 28 feet of white stuff to dig a grave. At the interment the pallbearers carried Kenny's remains down a series of snow steps cut into the side of the ice-walled mausoleum.

Twelve years earlier another priest, Father O'Donnell, had been the object of a snow-shoe sermon in the Downieville Mountain Messenger. Its report:

"Father O'Donnell, returning from Forest City lost one of his shoes near Mt. Vernon Ravine. He worked down the ridge on one shoe and got down the canyon. He then walked into Downieville. It's a good plan for those not accustomed to snow-shoes to tie them to the waist with a stout string."

Among the last of God's snow-shoeing messengers was Patrick Henry Willis, believed to have been the

The Rev. Patrick Henry Willis, who began and ended his career on snow-shoes in the Sierra Nevada.

only preacher of any denomination to begin and end his career in the high Sierra. Not only did the Rev. Mr. Willis approve of winter sports, but he himself often used snow-shoes in carrying his message to out-of-the-way points in the snow country. Stationed at Quincy from 1902 to 1909, his introduction to snow-shoes was at Johnsville, where he was called to perform a marriage ceremony in 1903.

"It was an unwritten law not to break any trails other than snow-shoe trails in Johnsville during the winter," the Rev. Mr. Willis once told the author as the cleric recalled how his host for the wedding party met him on snow-shoes. His Johnsville and other snow-shoe trips in the early part of the twentieth century left an indelible impression upon the Rev. Mr. Willis. So much so, in fact, that when modern-day skiing became fashionable in other regions of the Sierra he took to his pulpit to express a forceful viewpoint.

"The sports themselves," he pointed out, "are clean, healthful, wholesome and commendable, and those participating must be sober and clear-headed.

"The greatest participation," he continued, "is usually on the Sabbath Day and many of those of the Catholic faith attend early mass. This is commendable and in marked contrast to most Protestants.

"The Sierra Nevada mountains, however, with their stately pines and rugged grandeur, should awaken sublime thoughts and bring one close to the heart of the Creator. We approve the spirit and the purpose of winter sports."

The Rev. Mr. Willis had the unusual distinction of having been chaplain for legislatures of two states. In 1907 he came out of the snow country to officiate over the California legislature in Sacramento, and 20 years later he received a similar appointment to the Nevada lawmaking body. He was born in Illinois and came West in 1879 to reside with an uncle, the Rev. F. N. Willis, a pioneer mountain missionary. Not until his middle thirties, in 1901, was he ordained. Then for 40 years he was a familiar figure throughout the mining camps and resort towns of the High Sierra.

The very first fraternal organization to flourish in the snow country was the Jefferson Lodge of the Masonic order of La Porte, formed on May 6, 1856, while the camp was still called Rabbit Creek. During the summer of 1858 a Methodist Episcopal Church was dedicated by the Rev. G. C. Pierce. This asset to community life was followed on Dec. 8, 1858 by the institution of the Alturas Lodge of Odd Fellows.

The claim has been made that the Alturas Snow-shoe Club was formed as an adjunct of the Alturas Odd Fellows, but no such record can be found in the lodge's old minute books, However, its charter members, of which there were seven, nine years later all became members of the snow-shoe club. They were Barnes, the club's vice-president; Alex H. Crew, the club's treasurer; James St. Clair Wilson, H. G. O. Drake, C. Lowry, Charles Serch and D. Gore.

Among other early social organizations in La Porte was a military company known as the Sierra County Blues. Formed in 1858, its captain was Creed Haymond, later the first snow-shoe club president. This was well before the time the La Porters were to abortively seek the formation of a new county to be known as Alturas and, that failing, then to become annexed to Plumas County. The Blues were noted for their patriotic fervor. There was much parading and field maneuvers during the summer months, while in the winter it was not unusual for the militia boys to snow-shoe in from miles around to answer roll calls. Captain Haymond had many troubles keeping his men in line, with the result being a decline in soldiering's popularity when snow-shoe stake racing became fashionable. Later just about all the militia men were to join the Alturas club.

While it has always been the Odd Fellows who have asserted that it was they who "loaned" the name Alturas to the snow-shoe circuit, a similar claim might well be put forth for the Royal Arch Masons who organized an Alturas Chapter at La Porte in 1864. Its charter members included Haymond, Crew, Quicksilver Charley Hendel, John Corbett and a host of other men who were to become better known for their activities in the organized snow-shoe racing world. By that date there were also numerous other fraternal organizations in all the mining camps, notably the Pilot Peak and Table Rock lodges of Odd Fellows, Masonic groups in St. Louis and Gibsonville, and, of course, the widespread honorable order of E Clampus Vitus.

There was much traveling about to lodge meetings during the winter months. Naturally all the members were snow-shoers. The record of that is found in a March 17, 1867, communication from the Pilot Peak Lodge notifying brothers in La Porte that its members would be coming across the snows to attend a convocation. The trip was made a few weeks following the first successful tournament of the Alturas snow-shoeists. The Pilot Peak brothers were joined by others from the neighboring communities.

Snow-shoe travel became especially popular with the fun-loving membership of E Clampus Vitus. It was during 1857 that the brazen serpent and other insignia of this order was originated and displayed in Sierra City. The order became the best loved of all benevolent groups in the California mining camps, with members familiarly know as Clampers. There were two sides to the order, the social and the beneficial. The social side enabled miners, merchants and others to have some kind of place to meet and have a good time outside of the saloons, which in those days were the principal places of social diversion. For a short time after its organization there was nothing but fun; then came the beneficial effort as tragic accidents and also death hit their neighbors. It was then that the Clampers began to get in some really good work. The rugged life in the Yuba River communities provided many such opportunities. Committees would be formed to raise money. In the case of a leaky church roof, a new one would be provided and poor folks who needed their wood bin or dentistry were other needs the Clampers filled.

During March of 1869 the Downieville Clampers snow-shoed to Howland Flat to attend the annual races of the Table Rock Snow-shoe Club and while in town assisted the local lodge in staging a Clampers initiation ceremony. The visitors took advantage of a right to change the ritual because they believed the one in vogue would not do justice to the rugged snow-shoe racers seeking admittance. The object of the initiation was that the candidates would not forget and that the rites would have a lasting impression. Their success has never been forgotten! The proceedings were rough, and some of the victims were forced to remain in town for several days in order to recover from their bumps and bruises. The initiation included a trip down the "Rocky Road to Dublin," a long ladder with a wheelbarrow; there was a "Royal Bumper,"

SNOW-SHOE RACES
four days of snow-shoe racing at
HOWLAND FLAT
under the auspices of the
TABLE ROCK SNOW-SHOE CLUB
commencing Monday, March 15, 1869
free dances during the week
and a GRAND BALL on St. Patrick s Night
at the Sierra Nevada Hotel
by order of E. Clampus Vitus
D.H. Osborn, President

An advertisement in the Downieville Mountain Messenger, and followed by Charley Hendel's glowing words in the news columns:

"Our Howland Flat neighbors will have next Monday week something great, under the auspices of the Table Rock Snow-shoe club, assisted by the Hon. Order of E. Clampus Vitus, by the permission granted by the G.R.G.I.A. of Mokelumne Hill, who will be present himself to see the fun - when and where, it is supposed everything will pass off satisfactorily. I hope the Downieville Clampers will turn out in a body for such a running they never saw. Their wants will no doubt be cared for. Send your candidates over and they will surely be elected to some office - if they are in good standing in the community. If they come over and can't ride fast dope, we will allow them to ride their poles. —"Quicksilver" (Charley Hendel) Downieville Mountain Messenger, March 4, 1869).

contrived from a barrel and a number of paddles, and none ever forgot the "King's Jump," which saw the unfortunates hoisted by a block and tackle, then dropped into a brimful tin wash tub.

Among those who had gone to Howland Flat to attend the races was William W. Kellogg, editor of the La Porte Union. He returned home a sadly tired man and, because he probably had been sworn to Clamper secrecy, blamed his aching bones on his snow sports adventuring. His report, which caused many chuckles among his contemporaries down in Downieville and Quincy, was soon in print, as follows:

"ON SNOW-SHOES—We have taken our first trip on snow-shoes- —We have tried riding on greased boards—we have had enough. On Sunday last, 'ye locale' started on snow-shoes on a trip to Howland Flat. Our shoes—those that Babcock made for us—were well doped with fast wax. (It proved too fast for us.) In going from La Porte to Slate Creek we had a jolly trip, but from that point to the Flat the trip proved too jolly for us. Yes, we are a badly used up man—lame, stiff and sore. We could hardly move on the day after our return. Snow-shoe trips are very nice, we know they are—but we prefer that someone else should make them hereafter. `Dope is King'—it triumphed over us—it was a complete triumph,

Newspapers flourished during the Snow-shoe Era. The Virginia City Territorial Enterprise was a vigorous daily, May 19, 1869. That year the little town of La Porte, Jan. 23, 1869, had its own newspaper. The Downieville Mountain Messenger was originally published in La Porte, which it left, March 19, 1864. La Porte was called "Rabbit Creek" until it was re-named Rabbit Town in 1855, then its present name in 1857.

at least we thought so when one of our shoes got away and ran down a steep hillside and we had to go after it. Fast dope is a good thing, providing you can stick to the shoes— we can't."

Kellogg's editorship of The La Porte Union was short and lively. It began during the summer of 1868 when he moved the Quincy Union, the ownership of which he had acquired two years previously, but moved it to the town on the banks of Rabbit Creek. The end came the following spring during the height of the Union's windy editorial campaign against Snow-shoe Thompson, the man from Silver Mountain. Sparks flew down the street and the newspaper plant went up in smoke—files and all. It was the last newspaper to be printed in the high country.

But journalism had flourished before Kellogg got there, dating back to the days when La Porte was called Rabbit Creek.

The La Porte Union's lusty record is to be found in county histories, and several of Editor Kellogg's flamboyant snow-shoe articles were saved for posterity by Quicksilver Charley Hendel. Kellogg was a native of Berkshire County, Massachusetts, and came to the Feather River as a young man in 1858. His was a varied career—miner, constable, justice of the peace, editor, lawyer and legislator. He was admitted to the bar after giving up the editorial pen; later he became an influential member of the state legislature. He was

This photo, postmarked April 9, 1909, was a postcard received by A. L. Mullen of Loyalton, California from his mother. She wrote; "Hello Al; Isn't this awful! The valley looked like this after our last storm. Aren't you glad you don't live here?". Richard Jenkins Collection.

pre-eminently a self-made man of the type so peculiar to the mining camps.

Men came into the snow country to establish themselves as editors of the mountain press. They reflected the spirit and sentiment of the mining communities' leading citizens who were above all patriotic, believed in law, order and the family hearthside.

The history of journalism in the snow country begins with the Gibsonville Herald, which made its first appearance during the Winter of 1853-54. The exact date is impossible to determine because of the destruction by fire of all its files. Of its first editor, a scribbler named Heade, little is known. He was succeeded in 1854 by Alfred Helm, who issued a supplementary edition known as the Gibsonville Herald and St. Louis News. The paper was delivered by special messenger in the towns surrounding Table Rock. In 1855, the paper was sold and removed to Rabbit Creek, where the title of Mountain Messenger was assumed and an illustrious newsmonger named Albert T. Dewey took over. The news game in Rabbit Creek was great and the new editor played his hand carefully in order to forestall any other printer from poking his nose into that booming town. He played a proper editorial role, in July 1857, when the town's name was changed to La Porte.

Soon the sheet had numerous correspondents throughout the region. They sent in colorful accounts of early-day snow-shoe stake racing. But the snow depths proved too great to suit Dewey, who in January, 1864, sold his interest to Jerome A. Vaughn. The new editor loved the snow. He also knew when opportunity knocked. On Feb. 21, 1864, a fire gutted the two newspaper establishments in Downieville and Editor Vaughn, four days later, was on his way down below. His types and press were packed by snow-shoe carriers.

Under Vaughn's editorship the Mountain Messenger became a repository for all the colorful stories of the Snow-shoe Era. True, the short-lived La Porte Union and other newspapers in the Yuba and Feather River canyons gave space to the snow-shoe correspondents—principally The Argus, The Plumas Standard and The Plumas National, all published in Quincy, and in later years The Sierra County Tribune of Forest City. None, however, gave so much emphasis to the sport's various ramifications as did The Downieville Mountain Messenger, which today is the oldest weekly newspaper of continuous publication in California.

It has been said that the press of an area echoes the spirit and sentiment of its people. It is considered a reflection of the age in which it publishes. This was unquestionably true of The Messenger during Vaughn's editorship and to some degree it has continued on down through the years. The snow-shoers' way of life intrigued Vaughn and under his supervision the columns of the newspaper were literally alive with snow-shoers. He was twenty- seven years old and a booming printer with plenty of adventuring behind him when he acquired the newspaper. Journeyman ability had been awarded him in 1856 in his home town of Chardon, Ohio. Next he worked several months in Chicago, and then removed to Waukegan, Illinois. He worked there on The Gazette for two years. He then boomed through the composing rooms of The Transcript in Peoria, Illinois; The Northron in Delavan, Wisconsin, and then stepped up a notch to become foreman of The Janesville Wisconsin Republican.

April 1, 1861, Vaughn trekked overland to California, arriving in Marysville, Aug. 10, 1861. For a short time he worked on the Marysville Express and Appeal. He then went to La Porte and later to Sacramento and San Francisco, pegging type in all those places. In the spring of 1963 he crossed the Sierra on horseback to Carson City, then on to Virginia City, Nevada. In the fall, he and the first snows of winter arrived together in old La Porte.

During the following years in Downieville, Vaughn became a discerning editor. In snow-shoeing he saw something extraordinary. This was years before the days of the FIS (Federation Internationale de Ski) and this country's National Ski Association. Feb. 8, 1868, a question was posed to the readers of The Mountain Messenger by Editor Vaughn. The answers have made this book possible.

Here is what Vaughn wanted to know:

"Who will be first to write a book on dope and scientific snow-shoeing, the most rapid means of locomotion known to man?

"Such a work might be made very interesting by a little enlargement on the subject in its different bearings. But in the absence of such a work it might be worth the time of some of our snow-shoeists to put their ideas in print to the columns of a newspaper. We are ready."

Directly below Vaughn's little problem appeared a notice that The Messenger was instituting its own "Northern Snow-shoe Express" to insure delivery of papers to all its many subscribers up in the snowbound mining camps and also to take care of job printing accounts throughout the region.

The response was terrific, as the old files testify, and it is to be hoped that the job printing department did as well. It was no wonder that when it came time to lay such old snow-shoe veterans as Benjamin Wilcox Barnes beneath the snows for keeps that there were those with the know-how necessary to cross the mountains to La Porte.

References:

The Downieville Mountain Messenger

The La Porte Union
Interviews by the author

> INVITATION
>
> *E. Clampus Vitus Ball*
> December 24, 1855
>
> American Hall, Rabbit Creek
>
> Managers:
> George A. Davis, Rabbit Creek, NGH
> R. Swett, Rabbit Creek, James
> Crawford, Spanish Flat, NGH W.H.
> Hackett, St. Louis, NGH Frank
> Schoonmaker, Gibsonville NGH
>
> *Mountain Messenger. Adapted from the Dr. Albert Shumate, collection.*

17

Tracks for The Iron Horse

central pacific tracks

The Argonauts who settled and then built up Western America never were of the retrograding class. Disappointment more often spurred them forward than it frightened them backward.

This trait in early-day California character was well illustrated on Friday, March 1, 1867, when a passenger train carrying Central Pacific Railroad officials became snowbound below Emigrant Gap while chugging up the grade to its newly established rail terminus at Cisco.

Peering into a storm-swept railway cut from a vantage point on top of a snow plow were Leland Stanford, the Central's president and former governor of the Golden State, and Charles Crocker, chief of construction for the gigantic roadbed and tracklaying project up ahead. For 10 days now, wild Sierra gales had halted the building of the transcontinental rail crossing of the Sierra.

The two rail tycoons were headed upward to spur their work crews into renewed activity, and damnation to flatlanders who proclaimed it was financial and engineering madness to continue laying iron rails across Donner Pass. Down below there were opponents of the Central Pacific, and they advocated other (and what were claimed to be almost snow-free) routes in the Feather River and Placerville regions as better suited to the wintry obstacles of a Sierra rail crossing.

Up in the snows, beyond Cisco, was a work force of some 5000 Chinese and 1000 whites. Chinese laborers were there whose ancestors had built The Great Wall.

The temporary predicament in which Stanford and Crocker found themselves was described by a newspaper correspondent who for 10 days had been struggling

Through toil-built mountain gates,
We come, O Sister States!
With Hymns of praise;
Where white Sierras rise,
Where green plains face the skies,
We grasp the victor's prize
To crown our days!
The wild, grand march is done!
The guarded ways are won
From sea to sea!
We see His Mighty Hand
Now clasp this iron band
To grace our matchless land,
Where all is free!

—

—From the official ode sung at Sacramento during the Celebration in Honor of the Completion of the Great National Railway across the Continent.

westward across the mountains. First he rode a mud wagon from Virginia City, then trekked on snow-shoes and a horse-drawn sleigh from Crystal Peak to Donner Lake and Cisco, finally footslogging down the tracks to the scene.

In his report to The Sacramento Union the scribe stated:

"We suddenly came in sight of Governor Stanford and Charles Crocker, bundled up in their large coats and comforters, mounted on top of the famous snow plow. There were four locomotives behind them, snow to the north, south, east and west, under, above and all around them. In fact they were snowed in, and could not get home to vote at the primary election."

Nevertheless, Stanford and Crocker knew then the experience gained by battling the elements in recent weeks was proof enough that the gigantic building project could and would be successfully punched through the mountains. Even as the two men surveyed the snowdrifts which trapped their train in the rail cut they knew that snowsheds had been proposed to conquer just such difficulties. Strongly covered cuts, a well-settled roadbed, the free employment of plows and frequent trains, such as surely would be required once the transcontinental connection had been completed would assure the operation of trains through the severest storms.

Four years of effort now lay behind the rail builders. Roadbed grading work had begun on Jan. 8, 1863, at Sacramento when Governor Stanford turned a few spadefuls of earth and Crocker told assembled citizens: "This is no idle ceremony."

The initial rails were laid eastward from Sacramento, Oct. 26, reaching Roseville, June 6, 1864. The railway terminus was extended to Colfax, Sept. 10, 1865, to Dutch Flat in July, 1866 and to Cisco, Nov. 9 of that year. The trains were eventually scheduled to reach Donner Summit in August, 1867.

To speed construction, however, the Central Pacific Company outdistanced its own track-laying race across the mountains with a wagon road. Further, an infant winter sports industry had blossomed as early as 1864 at Donner Lake.

Leland Stanford

This came about through a subsidiary company organized to construct the Dutch Flat and Donner Lake Wagon Road. This staging route, completed to assure access to rail right-of-way work camps, immediately rivaled the Henness Pass and Placerville routes across the Sierra. The new route attracted much of the traffic to and from the silver mines of Nevada.

Eventual completion of the Central Pacific to Truckee in 1868 saw the company abandon its wagon road. Travel over the other two stage and freighting routes was ended before long.

To man stations along its mountain roadway, Central Pacific officials sought out the services of expert mountaineers, including the now-famed snow-shoers of Plumas and Sierra Counties.

Each intermediate station and the terminals for winter travel between Cisco and Crystal Peak, Nevada, had their consignment of snow-shoes, sleighs and horses all manned by crews expert at battling through the fleecy white.

During January, 1866, a full year before the rails reached the snowline of the western slope, the stage road agents were welcoming valley folk who came over the Sierra for skating at the lakes and ponds around Donner Lake. The press reported, "the Norwegian snow-shoes, or rather skates, are well adapted to locomotion on the snow. They are 6-8 feet long, bar shaped like the runner of a sled, four inches wide and two inches thick in the middle and grooved underneath."

A moot question was whether snow-shoes had originated from the Henness Pass or Placerville regions. Coals were tossed on that budding controversy when The Downieville Mountain Messenger in February, 1866, noted the death of a claimant to the honor:

"Another fatal accident occurred in the Monitor claim at Saw pit by which Charles Nelson, the somewhat famous snow-shoe expressman who is said to have first introduced the snow-shoes now in use in the Sierra, lost his life."

The Central Pacific's pilot snowplow is pushed by seven woodburning locomotives to clear the Central Pacific 1867 snow blockade west of Truckee. Southern Pacific Historical Photo.

And The Sacramento Union queried: "Where is Thompson?" Where snow-shoe tracks blazed the way, so went the mountain stage roads, followed by the iron rails and then the iron horse.

As Leland Stanford and Charles Crocker faced the blizzard of 1867, it was evident that they would no more turn back than had hardy pioneers in other regions. They could recall the Gold Lake excitement, which in 1850 led so many miners into the then

Central Pacific track clearing crew stands aside at Blue Canyon, between Baxter and Nyack, west of Donner Pass, to get out of the way of a passing freight. Southern Pacific Historical Photo.

unexplored snowy regions of Plumas and Sierra Counties; then the following rushes to the Kern River and Fraser River, Mono Lake and Washoe and how each rush, even those disastrous to the hopes of many, had resulted in the establishment of new communities and settlements newly etched in the history of California.

There had been a holiday spirit pervading the mountains as the Central Pacific Railroad's historic construction entered the winter months of 1867.

Huge sledges were constructed at Cisco well in advance of the first snowfall and, during fair weather in January and early February, the Sierra's gargantuan freighting operation fought through the snow-covered mountains with countless tons of construction material. Three locomotives, four railway cars, materials for 40 miles of track and other supplies were hauled across the snow to a supply depot at Donner Lake. Loads were dropped off at in-between wagon road stations in need of supplies.

Central Pacific crew struck a pose on this trestle to show off their ingenious new "Rotary No. 2." Soon replacing the old hard to push "Pilot Plow," the new rotaries were able to keep the tracks clear of deeper snow in heavier storms pushed by fewer locomotives. Southern Pacific Historical Photo.

Meadow Lake, on the wagon road between Cisco and Donner Pass, was a busy place in the winter of 1867-68, when the following letter was written by Clarence M. Wooster:

"We expected to move out before winter, but an early storm spread two feet of snow over the ground. This was quickly followed by a second storm and we were trapped. Father improvised snow-shoes for the horses and safely got them out, together with the men. At Cisco, fodder for the animals became a problem, for freight trains had been impeded by the unexpected and heavy fall of snow. It became necessary to rent a cabin at Cisco and convert it into a stable. Before the winter was over our horses for a time subsisted on corn meal. Just prior to the storm, the exodus from Meadow Lake City had begun. By wagons, over the newly made road, by mule trains and afoot, the surplus population moved to Cisco, where they could find transportation out by train. The early storm, however, prevented many whose luggage was of importance from going. So it is estimated that three thousand people wintered at Meadow Lake City in 1868.

"One day mother flirted with a pair of snow-shoes. A slight crust covered the snow. She had been an expert ice skater in Wisconsin. Starting down a short steep grade she exclaimed, 'isn't this delightful?' Then over she went, and hoops and woman, snow-shoes and pole, were all piled up together at the bottom of the hill. I never saw mother on skis after that, though for five months no one could move out without them.

"Storms were quite frequent. In the little valley where our house stood the accumulated snowfall measured fourteen feet. Snow-shoe Thompson passed us daily, carrying the mail between Meadow Lake City and Cisco. After each storm he would carefully make his track in the soft

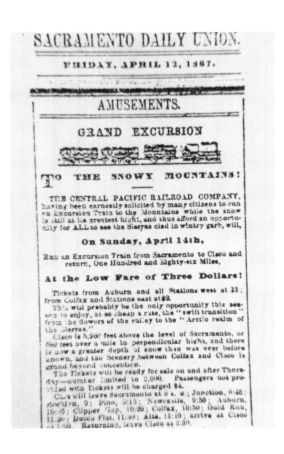

snow. Starting at the top of Red Mountain he would glide along the mountainside on a consistent grade. A frosty night would freeze the track, which would thereafter guide him as the steel rails do the locomotive. We would watch him sail down his four-mile course at great speed, cross the ice frozen river, throw our mail toward the house, and glide out of sight, up and over a hill, by the momentum gathered in the three-mile descent. A three-mile toboggan slide was a burning temptation which we resisted for some time; but finally the lure outran discretion and we trespassed on sacred grounds. We skied alongside of Thompson's track half a mile up the grade. His track was about six inches deep and frozen hard. We sat ourselves in it, squatted, holding our poles at right angles, and let loose of our tree-limb hold. We shot out like rockets. The skis held to the track, but the three kids went tumbling down a steep mountain side, head over heels, a hundred yards before sprawling out limp and seemingly boneless, scratched, and bleeding more or less profusely.

"Several days later while we were out on the opposite side of the great canyon and near Thompson's trail, he came along and called us over to where he was. Three kids then and there received a spanking which they never forgot. We did not repeat that adventure.

"Our stock of food ran low. Mother tore up a mattress from which she made three knapsacks, and we were dispatched to Meadow Lake City for flour, bacon, potatoes and other lesser needs. There was no city. Its people were celebrating a ski race, half a mile down a mountain side and out over the frozen snow-covered lake. We stopped awhile to partake of the excitement and then hunted for the town. It was entirely obliterated."

Mountain folk exulted as the maximum effort got under way to complete the trackway. Those who had established provision supply depots, small hostelries and otherwise, which were catering to the work crews, were particularly happy. The stage coach route from Dutch Flat to Donner Lake was described as "romantic and rather pleasant" for winter travel.

From Pollard's Donner Lake House a correspondent noted:

GRAND EXCURSION TO THE SNOWY MOUNTAINS!
Sunday, April 14

The Central Pacific Railroad Company, having been earnestly solicited by many citizens to run an Excursion Train to the Mountains while the snow is still at its greatest height, and thus afford an opportunity for ALL to see the Sierras clad in wintry garb, will run an excursion train from Sacramento to Cisco. This will probably be the only opportunity this season to enjoy the 'swift transition' from the flowers in the valley to the 'Arctic realm of the Sierras.' There is greater depth of snow than was ever before known, and the scenery between Colfax and Cisco is grand beyond conception.

Charles Crocker signed this Sacramento Union ad for an April 14, 1868 "grand Excursion to the Snowy Mountains." The train to the snow at Cisco Grove reached 5,960 feet "above Sacramento level," featured stops at Rocklin, Newcastle, Auburn, Clipper Gap, Colfax, Gold Run, Dutch Flat and Alta.

"I think it a pity that the citizens of Sacramento should remain in ignorance of the pleasant source of amusement within such easy reach of them as a 12 or 15 hours ride. Nature finds it hard, anywhere in your valleys, to make plenty of parks, but the skating pond could not be accomplished anywhere within our state borders so well as at this lovely lake. It is a fact that all these amusements, delightful and healthful, await you here. You may suppose the weather to be very cold here, so near Tip Top, but it is not as cold as in New York at this season, and residents from the western prairies express their astonishment at the comparative mildness of the winter atmosphere. The California Company's stages pass us each way daily on their way to and from Virginia City and, by the way, they are generally filled to capacity. We also have a telegraph office; so, you see, that when visitors come here from the outside world they do not find themselves cut off from all chances of communication with it. There is also a blacksmith shop, where skates can be made or repaired; but it would be advisable for skaters to bring their skates and not

Increasing freight traffic over the Donner Pass required bigger and bigger engines. The huge engine above, shown at Truckee in 1936, had the engineer's cab in front so he would not have to endure the smoke that belched out as the train went through snowsheds and tunnels.

wait to have them made here."

As the snow depth increased, another winter sports devotee advocated:

"Get out of the mud! Mount up to the hills, where all nature is clothed in spotless white; where we daily tramp glittering diamonds under our feet. Come up in the pure air and dwell nearer Heaven. Only think of going a mile-a-minute on snow-shoes, ye foot-pads in the valley mud! Your railroads are at a discount, your horse and buggy going only a mile in four minutes, could not be thought of, while remembrance of your slow steamboats and horse-cars make me nervous, Should the snow continue to fall, all humanity will be on snow-shoes."

the winter of 1867

Soon the winter of 1867-68 turned into one of the most severe ever recorded in the Sierra.

January 23, The Nevada City Transcript reported that at Meadow Lake, near Cisco, "the snow is about 10 feet on a level, and the traveling, except on snow-shoes, is rough."

Next The Quincy Union, no friend of the Central Pacific route, was out in print claiming that at no time during the winter to date had there been more than two feet of snow at Beckwourth Pass, and it asked: "What do the Dutch Flat railroaders think of that?"

Then The Sacramento Union retorted:

Horse-drawn sleighs provided transportation at Tahoe City, Truckee and Donner Lake during the late 1800s and early 1900s. The above photo was taken at Lake Tahoe.

Cisco Grove the "jumping off point," for the Central Pacific Railroad during construction reaching over Donner Pass to Truckee and the place suggested by Snow-shoe Thompson for a major ski meet, was to become Northern California's first developed ski area. Above, 1938 photo shows Clint Mason at the information desk of the Auburn Ski Club's jump prepared for the California Class A jumping meet. The railroad right-of-way can be seen at the top of the jump in-run.

"Why, we think the Pacific Railroad cars pass through deeper snow than that every day without the least trouble or inconvenience. They haven't enough snow at Beckwourth Pass to make it an object for a person living in the valley to go and see it, providing a rail car should ever run there."

Then came a January 26 report in The Downieville Mountain Messenger:

"The biggest snowstorm of the season is upon us. There is more snow now in Downieville than we have ever seen before and the storm continues. The snow on the summits must be very deep, and should it continue falling a day or two longer at the same rate, they will have to tunnel through it from one side of the street to the other in many of the upper towns, as they found necessary to do several winters some years ago. There will be fine snow-shoeing on the summits, as the ground will be covered, leaving a clean track for lightning dope."

Scores of lives were to be lost in the weeks which followed as snow slides roared down from the upper battlements of the High Sierra in Plumas, Sierra, Butte, Alpine, Placer and Nevada Counties. Records of that winter kept by Central Pacific construction engineers show 44 snowfalls, varying in length from a short squall to a two-week blizzard and in depth from a quarter inch to 10 feet. These storms struck at the work force spread over 20 miles between Cisco and Donner Summit where men were laboring under the harsh conditions of distance, bitter cold, snow blockades and granite cliffs.

An epic battle was won in the two weeks following February 9, 1868, when railroad officials announced they had begun "to haul over the snow, 3000 tons of rails, sufficient to lay 30 miles of track from Cisco to the Truckee River to be laid early in the spring, there now being a large force grading the road along the river. When the heavy work at the Summit is completed, this portion of the line being ready, they will be able to operate the road immediately to Crystal Peak, Nevada."

The fight to hole through Donner Summit, 7100 feet above sea level, was described in the Feb. 11, 1868, edition of The Virginia City Territorial Enterprise:

"Although the rocky peaks about the Summit, towering 1100 feet above Donner Lake are heaped and rounded up with snow, the work of driving forward the various tunnels necessary at this point of the railroad goes bravely on. At both ends of the main, or longest tunnel, gangs of Chinamen are to be seen like ants swarming in and out, while in the middle ground are to be seen, shooting into the cold thin air, great globular masses of more than snow-white vapor from the escape-pipe of a winter-smothered shaft house. Here the shaft has reached the level of the grade, and the workmen have faced about and are driving either way. Thus upon four faces at once is work going on in the great tunnel. The progress made, we are informed, is about six feet per week. Further east and far out around the southern shore of Donner Lake, away up on the walls of the almost perpendicular cliff, two or three shorter tunnels are being bored and battered through the hard gray granite projection out across the line of the intended track. Looking far up we saw in one place, standing upon a naked pinnacle of granite, what seemed to stand in the direct path of the engine in its approach and entrance into a great column or peak nearby, what at first appeared two or three ravens, pecking spitefully at each other, each pecking in turn. Looking more narrowly at the little black figures, so grotesquely bobbing upon the spire of rock, and aided by a background of snow, we at length made out three Chinamen, one holding a drill, and the other two striking it. In other places the same black specks were moving along the steep walls of the cliff, all boring and hammering away at tunnels in the great granite domes. How cold it looked away up there, how far away and lost in the dreary glitter of the frosted snow!"

Avalanches thundered at Donner Summit as the

aftermath of a huge snowfall which blockaded the rail route in early March. Travel was limited to hardy men on snow-shoes. The riders of the flying longboards brought word of snow terrors when telegraph lines were knocked down. Three construction workers lost their lives when a work camp at Donner Summit was buried by a snow slide, while at another point near the tunnel workings, the frozen bodies of 15 Chinese were dug out.

The Dutch Flat Enquirer reported:

" James Harvey Strobridge, famous as Crocker's construction supervisor, was on the ground with a party of men as soon as he got news of the catastrophe, and labored with great energy and determination until he succeeded in liberating the buried party at the camp log house. Ten were rescued without being injured."

On March 12, 1868 the summit's greatest snow depth in many years was reported by an agent for Wells Fargo and Co: "Bringing treasure through the snowbound Sierra in a sleigh, the white mantle is said to be within two feet of the telegraph wires, making it 22 feet deep on the level."

The St. Patrick's Day snow-shoe racing championship, with railway gandy dancers and miners competing for prizes, was won by S.C. Chambers in 17 seconds down a 1500-foot straightaway near Meadow Lake, while a Miss Clara Head took the feminine event in 25 seconds to win a gold buckle engraved with snow-shoes. There was dancing, too, at the Excelsior House for the beauty and chivalry of the snowbound Meadow Lake district.

Editors of the Meadow Lake Sun, awed by the snowpack, reported:

"Chimneys are of no use now. The last seen was some weeks since; nothing has been seen of it since that time, however, although diligent search has been made for it, by sinking a shaft down 35 feet, and then fishing for it with a 30-foot pole. It could not be found, though it stood 30 feet above the roof of the building, which, by the way, is three stories high. The shaft to China has likewise been abandoned. Instead of sinking inclines we have all run tunnels to the center of the Plaza, at which point we have a large iron cylinder, 30 feet in diameter and 500 feet high, with a self- moving lid on top; it is through this we ascend when we are so foolish as to get up in the morning. We intend to break a road from here to Cisco on Monday next."

Rail travel resumed to Cisco, March 20, 1868 with a roving writer soon advising the Marysville Appeal:

"The cars have been coming in laden with freight for over the mountains, Virginia City, Austin and other places. We have had our equinoctial storm, which was quite light. Only about a foot of snow has fallen during the past week. Snow all told now measures about 12 feet. Freight is very abundant, and teams scarce and in great demand. The sleigh road to Crystal Peak is in fine order for sleighing. Freight to Virginia City readily commands from 2-3/4 to 3 cents, which is a very fair price considering the condition of the roads. Stages are on runners a distance of 50 miles from Cisco, the balance of the way using their coaches. With regard to the opening of the railroad with the big plows, they have got them so they are a perfect success. They put behind the snowplows three locomotives, then get a good start, plunge into it, and they will elevate it some 14 to 15 feet, doing more in execution that 300 Chinamen shoveling. I think with the aid of the snowplows there will be no more delay in running the trains."

In Virginia City, The Territorial Enterprise echoed:

"The Central Pacific Railroad is drawing gradually toward this side of the mountains. Many have spoken derisively of the route by which the road is coming as 'the Dutch Flat trail.' Men will differ about such a thing. Many of the troubles which beset the working of the road are now in process of being remedied. Gigantic work is doing, on the surface, against distance and altitude, and in the face of nature's strong barriers, the ribs of the towering mountain range. Let us hope the worst is over, and, while its managers work away, let us wish all speed to the Central Pacific Railroad!"

Meanwhile spring was smiling on the valley of the Sacramento. The plains were painted with wild flowers; pale green young wheat was bursting out, and orchards frothed blooms with promise of fruit. Yet valley residents easily could view the distant white summit of the Sierra, etched against clear blue. They knew that only five hours away over Central Pacific rails there were snowclad hills down which snowshoers skimmed. Up there in the heights, sleigh-bells were ringing as muffled people sped along beaten roads walled with snow towering over 10 feet high.

If horrendous reports of snow 100 and more feet deep in Meadow Lake did not quite square with the truth, at least there was other documentation of what had been a long winter, charged with frequent and prolonged storms. Rail officials agreed that the first great fall of snow had been readily managed, even though trains had been checked running through to Cisco by washouts below the wintry belt. However, before the line had been fully restored, other storms broke upon the mountains, and from that time until mid-March the snowplows had been kept in continuous service. It had been a costly experience, which the rail builders decided to eliminate by covering track with snowsheds rather than widening troublesome railway cuts.

April 9, 1868, the valley towns were plastered with broadsides. The newspapers carried advertisements which proclaimed:
GRAND EXCURSION TO THE
SNOWY MOUNTAINS!
Sunday, April 14
Over the signature of Superintendent Charles Crocker it was announced that,

"The Central Pacific Railroad Company, having been earnestly solicited by many citizens to run an Excursion Train to the Mountains while the snow is still at its greatest height, and thus afford an opportunity for ALL to see the Sierras clad in wintry garb, will run an excursion train from Sacramento to Cisco. This will probably be the only opportunity this season to enjoy the `swift transition' from the flowers in the valley to the `Arctic realm of the Sierras.' There is greater depth of snow than was ever before known, and the scenery between Colfax and Cisco is grand beyond compare."

Cisco, a fine site for snow-shoe competition, had been reached by the public and would later become a winter sports mecca.

Southern Pacific snow-train unloads at Truckee, 1931.

On May 8, 1869, free trains were provided celebrators by the Central Pacific Railroad. Thousands poured into Sacramento for the "Celebration in Honor of the Completion of the Great National Railway Across the Continent." Two days later Leland Stanford drove the

Golden Spike at Promontory Point to officially note that the, "guarded ways were won."

At Sacramento, the signal that announced that the last rail had been laid was given by a cannon shot from Union Boy and a simultaneous blast from 23 locomotives on the levee and the ringing of all the bells in town. The locomotive, "Governor Stanford," whistling in its position of honor, was the first to run on the Central Pacific Railroad. The clamor lasted 15 minutes in advance of the grand procession. The day was just warm enough to delight a Sacramentan, but was a little too warm for men who had trooped down from the mining camps.

Governor H. H. Haight's address recalled how men had conquered the mountains:

"In the winter of 1849-50 the streets of San Francisco thronged with miners, driven there for shelter from the inclemency of the season. The prevailing style of dress was a flannel shirt in lieu of a coat, and in addition to the ordinary nether garments, a pair of boots purchased at the moderate price of six ounces, or $96 in gold dust. Most of the large rooms of the city were used as gambling saloons, with the accessories of bands of music and well-stocked bars. Here day after day were to be seen dense crowds of men of all nationalities and races, ignoring all distinctions of race. A lady was sufficiently rare to cause a street full of men to stop and turn to look at one passing.

"The advent of spring dispersed the crowds of miners throughout the mountains, and produced a stagnation for the time in San Francisco. The public gaming house was soon succeeded by the school house, the hospital and the sanctuary. The common law and a code of well-considered statutes superseded the vague, uncertain and strange rules of Mexican and civil law. The other feature of our presence is the great Continental Railroad, the completion of which we are met to celebrate today.

The "pilot plow," created in 1867 to battle the heavy snows that blocked Central Pacific trains in the Sierra Nevada, was pushed by as many as seven balloon-stacked locomotives. Above, wearing a frock coat, stands Leland Stanford. Southern Pacific Historical Photo.

"It was an arduous undertaking from first to last. The progress, at the outset, was slow through the foothills and up the mountain slopes of the Sierra. Hills were cut through and canyons bridged, until after about four years of labor the iron track reached the solid granite of the summit. A tunnel of 1650 feet, through granite rock, involved enormous expense and labor, and caused a delay of a year. The mountain was finally pierced, and the iron track started but a little more than a twelve-month ago, ran its race toward the Rocky Mountains. That the rapidity of its construction since that time has been a marvel, and that all anticipations have been exceeded, is notorious.

"The day is at hand when a more splendid civilization than any which has preceded it will arise upon these distant shores. A vast population will pour into this Canaan of the New World.

"Tourists will be attracted by the most sublime scenery on the continent, and thousands will come to repair physical constitutions racked by the extremes of climate, the inclement air, and the miasma of the states east of the mountains."

During a collation in the assembly of the state legislature," Gov. Haight concluded:

"Charles Crocker gloated over his triumph:
'I recollect when Governor Stanford pitched a few shovelsful of dirt, at the commencement of this work, what sneers and jibes and taunts were made use of at that time. I told you at the time the work had commenced, that it never would stop until it was completed, and now we have a railroad that spans the continent, and today some of the men who traduced the managers of this enterprise, the very men who said it was a "Dutch Flat swindle," are now pushing to the front. Now I wish to say to you, that we will put the rates down so as to fill this country full of people. Just as fast as we can get the cars built we will bring them here and fill your valleys and mountains full of people.'"

Soberly reflective in nature were talks given during a collation of the Sacramento Pioneers' Society. Hospitality to the invited guests was dispensed with an easy grace and without formality under the direction of

Mr. & Mrs D..L. Bliss, pioneer Glenbrook family that built Tahoe Tavern and narrow gauge railroad from Truckee to Tahoe City. Bill Bliss Collection.

Old sleigh takes off at Tahoe Tavern, the Lake Tahoe resort built by D.L. Bliss of Glenbrook. WBB. photo.

Southern Pacific "double header" standard gauge snow train steaming along the Truckee River toward Tahoe City in 1932. The broad gauge track was laid in 1927 after D.L. Bliss leased the short line, originally narrow gauge, to the SP for 99 years. The price was one silver dollar. WBB photo.

James McClatchy, president of the organization and publisher of the Sacramento Bee. There were representatives from San Francisco, Stockton and Marysville pioneer associations, and pioneers from many other counties. Following dinner, President McClatchy in a short talk referred to "the grandest poem of modern times—the great Pacific Railway" as having been just presented to Sacramento and the nation.

Among those present were old soldiers of Stevenson's Regiment, of Scott's and Taylor's Divisions of the Army of Mexico, Californians whose residence dated far back of 1849 and even a member of the ill-fated Donner Party of 1847.

The Donner Party veteran, William Murphy of Marysville, in a short speech alluded to the fearful sufferings of that party of lost emigrants in the deep snows of Donner Lake. "Lost," he said, "on ground within sight of the great highway which now defies all obstacles of nature to baffle the energy of man."

The Golden Spike ceremony out in the Utah desert was somewhat anti-climactic for Sacramentans still hoarse and deaf as the aftermath of their own celebration. Nevertheless, the final events were made interesting by the presence of Ex-Governor Stanford, dignitaries of the Union Pacific Railroad and several hundred roughly dressed spectators drawn from the work crews of both lines.

Warm and earnest greetings were exchanged between officials as they fondled a ceremonial railroad tie of polished laurel wood from California. An orator told how the golden spike was to "become a part of the great highway which is about to unite California in close fellowship with the sister states of the Atlantic."

The cowcatchers on the locomotives of the two railroads touched. Champagne was passed from one engine to the other. The locomotives in turn crossed the magic tie and the union was consummated forever. Behind Leland Stanford stretched 690 miles of rail, over the Sierra to Sacramento, and farther west, another 126 miles of track to San Francisco. Up ahead were 16 new cars, first-class passenger units, for the Central Pacific. They stood ready to be coupled to Stanford's train for the long haul home. Soon the cars rolled westward.

Just because the transcontinental railroad had been completed did not mean that the inevitable snows destined to hit the Sierra crossing were to be conquered for all time. Things temporarily looked up June 3, 1869, when the first of the Central Pacific's Silver Palace Sleeping Cars arrived in Sacramento from the East. It was noted that the sleeper was "slightly wider than ordinary passenger cars, and in passing through the snow sheds of the mountains, rubbed against the timbers occasionally."

Soon, however, there were challengers for the Silver Palace. Both put on a grand appearance June 10. They were the sleeping car "Pennsylvania," accompanied by a diner named "International." They rolled into Sacramento with their builder aboard. He was the famous George M. Pullman. The newspapers reported, perhaps ominously, that Pullman wanted to bring a larger sleeper, but "it was unable to get through the snow sheds."

Ice from Donner Summit was being advertised for sale at three cents the pound under an admonition to "patronize home industry."

Two mountain resorts; the Summit Hotel at "the summit of the Sierra on the railroad," and the Lake House, promising "saddle horses always on hand and bathing rooms fitted up on the Donner Lake shore," published announcements that they were ready for the tourist trade. Rail passengers going east were being importuned to supply themselves with "photographic views of the sublime scenery along the route." All this may have looked like handwriting on the wall to a Sacramentan named Jim Farris. He was advertising to sell a pack train of 12 mules for a good price.

Officials of the Central Pacific, soon to be called the Southern Pacific, now sought to match the forces of King Snow by completing a gigantic snow shed construction program along the rails through the Sierra. The shed-building project, begun in 1867, was to see 40 miles of track covered by 1873. The sheds would stretch in dismal and smoke-blackened incongruity from just west of Truckee to west of Blue Canyon.

The earliest sheds were similar to a house, with steeply pitched, peaked roofs. Logs for posts and braces were felled in the forest alongside the tracks. Neighboring sawmills supplied the timbers for the roofs and sides. The posts and braces were buried deep in the ground to provide rigid alignment.

However, the design of the sheds proved impractical. The posts rotted and offered little resistance when deep snows threatened to force the sheds out of line.

Snow sheds soon were designed with flat roofs. The posts and braces were anchored in concrete. Clearances for passing trains were increased from 17 to 23 feet. In many places concrete walls were poured on the sides facing the mountains.

Snow-fighting outside the sheds was accomplished initially by push plows and the strong arms of human shovelers. Then came the cyclone plow, a contraption resembling a huge screw driver. Neither of these methods was efficient. Next the chore was handed over to the newly-designed rotary plows, introduced in the later winter of 1890. These were an immediate success and, with improvements, have battled effectively the snows of all but the severest storms.

As the efficiency of snow-fighting equipment increased, there seemed to be little need for sheds over the tracks in many locations. The original 40 miles had been reduced a few miles even before 1890 when the route became snow blockaded for 10 days. When double-tracking was extended from Colfax to Blue Canyon in 1913, the sheds in that district were abandoned. By the time double-tracking had resumed in 1924 there was slightly less than 30 miles of shed remaining.

References:

The Nevada City Transcript
The Quincy Union
The Sacramento Union
The Downieville Mountain Messenger
Central Pacific Railroad records
The Virginia City Territorial Enterprise
The Marysville Appeal
The Dutch Flat Enquirer
Wells Fargo & Co. Reports
The Meadow Lake Sun
The Sacramento Bee

Truckee became a jumping-off point for big rigs in 1936, providing stiff competition with the railroad, except during big snow storms. W.B.B. photo.

18

A Ski Center In The Sierra

Truckee, which had become a thriving rail center, was destined to become the hub of the greatest winter sports recreation mecca in the western hemisphere. Boomtimes during the next three decades would see rail lines spread through the mountains to haul away lumber and mine riches.

It was difficult during January of 1869 to find a more beautiful Sierra Nevada winter setting than the railroad town of Truckee on a moonlit night. Cheery lights gleamed from houses scattered about the snowy landscape. They twinkled against the snowclad backdrop. Beyond stood the forested hills, thick with dark pines, shouldering their frozen winter dress.

Sleighs flashed by with bells jingling. They passed mountain men, who poled along on snow-shoes we now call skis. Children gave life to the winter scene as they slid on sled-like contrivances known as "Yankee jumpers."

In 1916, Truckee was flourishing as a winter sports center. This view of Downtown Truckee is taken from a vantage point on the railroad tracks looking north across the snow covered Main Street, featuring a skating rink, at right. Carson White Collection.

Central Pacific Railroad trains chugged through Truckee, as they tried to maintain regular schedules in their continued fight through the drifts along the Donner Summit route.

The rails linked Sacramento to a newly established terminal several hundred miles eastward into Nevada. Beyond this point there were 12,000 Chinese laborers and 1,000 white artisans laboring under the supervision of construction boss James Harvey Strobridge.

Strobridge's driving genius had conquered the mountains. He continued pushing tracklaying crews toward Promontory Point, Utah. At that remote point his reaching rails would link up with those of the Union Pacific's onrushing work crews. This heralded meeting, now known by every schoolboy, would signal completion of the nation's first transcontinental railroad. The redoubtable barriers had been conquered. The era of the iron horse had peaked. Meanwhile the interminable, noisy engine serenades echoed through Truckee's rail yards.

The clanking engines heralded the end of the horse-drawn stage and freight service over the Sierra. Their heyday was fast drawing to a close.

Iron horses puffed and whistled. Red and green lights twinkled as strings of cars clicked over the new rails. Snow-plows, timber-laden cars, freight trains, passenger expresses pulled and shoved. The smooth rail had superseded crude, rough highways. Comfortable cars had replaced over-crowded sleighs.

Abandonment of the Dutch Flat and Donner Lake Road came first. Soon the Placerville and Henness Pass routes were doomed. The toll roads of winter were quickly reverting to snow-shoe trails.

Mountain roadways, however, would have another day. Economic change, heralded by engine whistles, soon spread throughout the land.

Truckee, which had become a thriving rail center, was destined to become the hub of the greatest winter sports recreation mecca in the western hemisphere. Boomtimes during the next three decades would see rail lines spread through the mountains to haul away lumber and mine riches.

Wherever the iron horse went, so would go the snow-shoe. Those slim spruce longboards had uses far beyond the ken of many flatland folk. The rails would be removed after the wealth had been freighted away. Then the snow-shoe tracks again would lead the way as highway builders faced the Sierra Nevada challenge.

The pattern of events during the winter of 1869 was to be paralleled many times in the years to come.

Perceptive Central Pacific officials encouraged metropolitan newspaper correspondents to ride their comfortable cars through the mountains that winter. Among the rest was one bylined "Ridinghood," a venturesome lady scribe from San Francisco. She penned, "Summit of the Sierra," "Snowbound at Lake Tahoe" and "Trip Over the Mountains." Her tales describe in glowing terms a region twice tabbed for the Winter Olympic games in 1932, but successfully in 1960.

Her adventures included her rescue from the Tahoe House by eight men using snow-shoes, horses and sleighs.

Ridinghood made one observation in common with the male journalists who followed her writing trail. She declared that the snow-sheds of the railroads were "frequent and provoking," because they shut out the magnificent mountain views. But, nevertheless, she discovered a snow-shed advantage by reporting that, "Sometimes on going though a long one, some mischievous young gentleman pretended kissing was going on; and as I sat beside a bashful young man, one of my Washoe friends had the audacity to tease us both pretty thoroughly by looking at me quizzically and declaring that the feather on my hat had really sustained serious mashing!"

Travel by train did not always have light moments. On Feb. 10, 1869, the daily westbound passenger train became snowbound four miles west of Truckee.

The efforts of several locomotives to move the stalled cars forward or backward proved unsuccessful. The passengers remained stranded for 36 hours without food until relieved by the arrival of men on snow-shoes. Then it was observed that the face of a lady was breaking out in small spots, mistakenly identified as smallpox. A panic immediately prevailed among the passengers. An earlier decision to stick with the cars rather than hike to Donner Summit was quickly rescinded. Now, even though the snow had been 50 feet deep, they would have essayed to quit the wilderness by some means. Soon the travelers started, following the trails cleared by snow shovelers. They trudged though snow sheds and tunnels. In the face of word that all trans-Sierra rail travel was blockaded, they advanced in three days to Emigrant Gap before again boarding the cars for Sacramento.

The storm demonstrated to a Virginia City Territorial Enterprise correspondent that, "from Truckee to Alta, the Central Pacific Railroad must be shedded, nearly every rod, to be rendered practical in the winter. Wherever the sheds are, two engines with a plow can clear the way. In other places, ten are inadequate to the work."

Clearing weather, Feb. 19, 1870 at Truckee, saw no less a personage than Snow-Shoe Thompson boarding the cars. His passing was noted by a puff story in The Alpine Chronicle:

"Our fellow citizen, J. A. Thompson, well-known as 'Snow-Shoe' has gone to La Porte to attend the snow-shoe races to come off next week. If the La Porte 'sports' allow our friend Thompson to take part in their runs, he will make them use their 'dope' pretty freely to enable them to cope with him.

"Thompson's leave-taking of Diamond Valley had been preceded by an event during clearing weather on Feb. 18, on the Comstock heralding the fact that the iron horse and rails were spreading though the mountains to temporarily replace sleighs and snow-shoes as a means of winter travel.

"A brief ceremony was staged as work began on the Virginia and Truckee Railroad which was to connect The Comstock with Reno and the transcontinental railroad. Of the ground-breaking event at American Flat, The Territorial Enterprise reported: 'There were no grand ceremonies on the occasion. The few gentlemen present took a nip or two of old bourbon, and with a few words, more in the shape of toasts than speeches, the thing was done. Some 40 men were set to work at once.'"

The destiny of the mountains was now spelled out by mountain editors in all directions. Snow-shoes would become known as skis and the High Sierra world-famous for its winter sports. The wild bursts of snow sports fever from which this was to spring were taking place at La Porte and alongside the tracks between Truckee and Cisco. Soon, however, the snow-shoeing thrills and controversies of 1869 would be forgotten by all but the mountain folk involved. For 10 years there had been an annual exodus from the gold camps to the silver camps. The names of those who struck it rich were being talked about by everyone.

benefactors

A group of millionaires would emerge from the era of mining, railroad and lumber promotions. Only a few, however, cared to remember the stirring times from which they had sprung. The vast majority were only too eager to forget the crude times before fortune smiled.

However three men never forgot the old times. John W. Mackay had snow-shoed across the mountains from Downieville during the initial rush to The Comstock where he became a Bonanza King.

E. H. "Lucky" Baldwin was rescued by Snow-shoe Thompson from the snows above Strawberry while traveling the Placerville route to the silver mines in 1860. The third, whose memory of the early days never faded, was Duane L. Bliss, a Gold Hill bank clerk, who was to become a Lake Tahoe timber and railroad baron.

The Mackay fortune endowed the University of Nevada with a school of mines, a science building and an athletic plant. From those structures emerged a world-famous snow science, an all-time ski great of the Sierra, and the national intercollegiate championship ski team of 1939.

Because of Baldwin's good luck, people now enjoy use of 1-1/2 miles of Lake Tahoe frontage known as Baldwin Beach, deeded to the U. S. Forest Service by grandchildren of the wily financier.

Thanks to Duane L. Bliss and family, the public has several heritages, including Bliss State Park and its 7,250 feet of Lake Tahoe shoreline at Rubicon Point. Bliss, the pioneer, built two narrow-gauge railroads to further his timbering operations above Glenbrook and near Echo Summit. Even as the lumber was being freighted and flumed to the Carson Valley and thence to Virginia City; for use in mining operations, Bliss had the foresight to issue orders that no tree under 15 inches in diameter should be cut. He believed that some day beautiful Lake Tahoe and its surrounding mountains would become one of the world's great year-round resorts. He furthered this vision at the turn of the century by building Tahoe Tavern, near Tahoe City. He then constructed a narrow-gauge railway line from Tahoe City to Truckee, linking this fashionable resort to the Southern Pacific Railroad where San Francisco travelers could change trains.

birth of a sports mecca

This was a stepping stone to the expansion of Sierra winter sports resorts. The boom saw Tahoe Tavern become a familiar name, which was bandied about to get the attention of the International Olympic Committee during the planning for the 1932 winter games and later for the 1960 Olympics at Squaw Valley.

During the early years the Bliss family established a ski area at Spooner's Summit, above Glenbrook. In 1946 they joined the Douglas County Ski Club of Nevada in launching the annual cross country ski event, memorializing Snow-shoe Thompson.

Further, it was the 1927 effort by Southern Pacific Railroad officials that provided impetus which was to rocket the High Sierra into world fame as a ski and snow sports center. They negotiated a 99-year lease with the Bliss family for the railroad right-of-way between Truckee and Tahoe City. The deal was sealed with a silver dollar. It was agreed that the narrow-gauged route was to be broad-gauged to permit switching cars from transcontinental trains onto the Lake Tahoe line. Simultaneously, the Bliss family transferred ownership of Tahoe Tavern to a private operating corporation. Thus Tahoe was opened for all-season vacation travel.

Truckee toboggan tower and slide, left, and ice palace, right, contained an indoor skating rink, drew winter sports lovers in 1900. John Crobett Collection.

Ever since 1890, Truckee had staged annual winter sports programs. In the early days they featured ice palaces, skating events, ski exhibitions a toboggan slide, a pullback and dog team races.

The winter of 1928 saw California businessmen thoroughly awakened to the economic expansion inherent in snow sports promotion. Near Tahoe Tavern lay a pine-sheltered slope to be named Olympic Hill. Here a toboggan run with a return saucer full of thrills was completed for the 1929 season. Construction of a 60-meter ski jump was begun for use in the winter of 1930. The project was supervised by Lars Haugen, seven times national ski champion from St. Paul, Minn.

Recognizing winter sports attractions, Yosemite, during the same period completed ice rinks for hockey and other skating events. They built a ski jump for exhibitions and blazed highland ski courses for mountaineers. The Yosemite expansion was made possible by the 1927 completion of an all-year highway to the snowline. The work was directed by Ernest Des Baillets, a French-Swiss snow sports expert, who had become known through his connection with the Snowbird Club of Lake Placid, N.Y. and the Chateau Frontenac, Quebec City, Canada.

The professional efforts of Haugen and Des Baillets

had the whole-hearted backing of the Winter Sports Committee of the California State Chamber of Commerce. San Francisco's Jerry Carpenter, who soon became known as the "Old Man of the Mountains," was the principal coordinator of development programs.

Leadership in the mountains was provided by Wilbur L. Maynard from Truckee, where he operated the Southern Pacific Hotel and for years had been the chief promoter of winter sports in the region; Wendell T. Robie, a lumber man from Auburn, soon to be tabbed as the first president of the California Ski Association; Henry Dorset, a strong supporter of winter sports at Lake Tahoe; Arnold Webber, a Placerville enthusiast who in 1935 was to conceive the original plan to memorialize Snow-Shoe Thompson; Husky Beresford of Mineral, resort owner, Mt. Lassen region; Ed Walker and Bill David, AAA manager, Reno, and Larry Evenson in the Mt. Shasta region.

The energies of this group was turned immediately toward obtaining the winter events of the 1932 Olympic games, which had been awarded to the United States. The summer events were scheduled for Los Angeles.

Right from the start, however, the planning group met stiff opposition from the state's sunshine lovers. Soon after the International Olympic Committee in 1929 turned down the state's bid at Lausanne, Switzerland, a report written by Carpenter stated:

"Our committee received tremendous opposition from the various community advertising organizations which felt the least mention of snow in "sunny" California would be detrimental to their investment of millions of advertising dollars in making this a proverbial land of sunshine and roses. That they succeeded was evinced by the fact that we could not make the International Olympic Committee believe that we had either the snow or the facilities to hold the games in California. The winter games went to Lake Placid, where they had to import the snow, while at the same time we had better than eight feet of snow on the hill at Lake Tahoe."

Anguished cries came from Wilbur Maynard, who asked:

"Are we not generously endowed by nature with terrain and other conditions? Ample transportation, housing and finance and other essentials? Yes, comes the answer, but, how was the world to know that these conditions existed? Where was our background for staging an event of this magnitude? Why had the world never been informed of this vast playground in such proximity to orange and almond blossoms? Where and why, indeed?"

Wendell T. Robie, founder of the Auburn Ski Club, pictured in 1930 when he was elected first president of the California Ski Association. He boldly championed the California Ski Association admission to the National Ski Association.

The California chamber's winter sports Committee then drove toward establishing the California Ski Association and affiliation with the National Ski Association of America. Then came the staging of the National Ski Jumping Championship at Lake Tahoe's Olympic Hill in the wake of the world-wide games at Lake Placid.

Sentiment of the mountain communities was echoed by Maynard when he told the organization meeting:

"We must face the problems that will come with good sportsmanship and good management. If we do not foster every angle of Old Man

Winter, we are remiss in our duty to those who need leadership, and to those whose health and general well-being, aside from the great commercial value, can be promoted most generally."

A clear account of that period was written by former State Sen. Bert A. Cassidy, Auburn, who was editor of the Truckee newspaper in the early 1900s when skiing was in its infancy.

Cassidy was a native of Chippewa Falls, Wisconsin, where skiing in the U.S. took root early. He came to Truckee in 1909 and bought the newspaper, then called the Truckee Republican.

"The Southern Pacific Hotel had closed for the winter," he wrote. "The community was starving and needed some new economic thrust to save it from extinction."

Cassidy dreamed of making it California's recreational winter playground for the nation.

"While skiing was available for the outdoor sportsman, the businessmen were soon united behind our winter sports program, because the more popular sport of whiskeying (pronounced whisskiing) brought visible returns in the form of a busy cash register. What appeared to be a dying town was soon one of the busiest beehives in winter economic life of eastern California. The sport was exhilarating."

celebrities

Charles McGlashen sparked winter sports development when he started the Truckee Winter Club in 1895. He wrote:

"With winter sports came the motion picture companies to give added publicity to the efforts of those trying to capitalize on one of the advantages nature had made available in such abundance.

"There was Hollywood's Hobart Bosworth, Tom Mix, Larry Peyton, Wallace Beery, Art Accord, William S. Hart, Ford Sterling, Mary Pickford, Lillian Gish and a galaxy of stars such as Hollywood alone could produce.

"With the movies came many authors. Fiction writers and scenario writers came to get color for their snow stories. Following the dog race of 1915 we ate trout with Jack London, his wife Charmia, John Johnson, Bill Brady and Martin Johnson, the big game hunter, who was cook on the 'Cruise of the Snark.' We dined in our home adjoining the printing

Jack London had dinner with winter sports pioneer and Truckee Republican editor and publisher, Bert A. Cassidy, who came to town in 1909. On that trip London managed to get in a little dog sledding, a skill he picked up in Alaska. W.B.B .photo.

office.

"Rex Beach, Peter B. Kyne, Jackson Gregory and many others in the writers' hall of fame visited Truckee in those thrilling days of Winter Sports.

"Then too, came newspaper men for color to make their stories more interesting.

"Occasionally when they hit a winter like that which trapped the Donner Party, their trains would be snowed in for several days. Special trains would bring food from Reno."

Cassidy detailed the growth of winter sports:

"Following World War One, interest was increasing in winter sports. Accessibility was ever becoming a more pressing problem of the state. Development of roads that could be plowed soon made it possible to give Californians the thing they sought—outdoor winter recreation.

"With the end of the war and the advent of Prohibition, the businessmen of Truckee again went through a reconditioning process. It was hard for the town to weather the storm of both Prohibition and a change in Snow Sports leadership," Cassidy wrote.

About this time Cassidy bought the Auburn Journal and Placer County Republican, both county seat newspapers.

Mrs. Wendell T. (Inez) Robie, on behalf of the California State Chamber of Commerce, handed Hans Haldorson, Auburn Ski Club representative, the statewide trophy, engraved: "Awarded to the club securing the greatest number of points over a five-year period in California Ski Association Championships, 1934-1938."

ski clubs

"It was in Auburn that the next big stride in Winter Sports was to be made, Cassidy continued.

"In 1929 the Auburn Ski Club was organized and this group of live-wire Californians soon gave the impetus to Winter Sports that has carried it to every city in the state. Thirty five Winter Sport or ski clubs have been formed reaching every section of the state. They all have their favorite ski grounds. Ready access to these snow areas, due to good highways and the snowplow crews of the State Division of Highways, bring them within a few hours ride of the most distant city."

Cassidy's enthusiasm for

The California State Chamber of Commerce trophy won by the Auburn Ski Club. The trophy is now exhibited at the Western America SkiSport Museum, Boreal Ridge.

winter sports propelled him into the California Senate.

"In the winter of 1930-31 the writer, who was then in the State Senate, arranged a party for the Legislature and State officials at the Cisco Grove grounds of the Auburn Ski Club. A caravan of Auburn citizens with autos took over 100 State officials and members of both Houses of the Legislature to the Winter Sport grounds on a Sunday in late January. Many had no first-hand knowledge of snow or snow sports before that visit. However, after that trip, it was an easy matter to obtain appropriations to develop roads into the snow areas of the Sierra and to obtain funds with which to clear the roads of snow, direct to ski hills in the state.

"In the winter of 1931-32 the National Ski Tournament was held at Lake Tahoe.

"After tentatively setting the date for the meet for Tahoe, state and national ski clubs officials began to fear the meet might fall through. Financing and housing arrangements seemed an insurmountable task. The writer was called in by state officials, the ski club and railroad executives. They asked if he could help save the prestige that would be lost for California winter sports if the meet was cancelled." He accepted the tournament directorship. A colorful meet was held that last week in February, 1932, at Lake Tahoe on a ski hill 6,200 feet up in the mountain vastness of the enchanting Lake Tahoe section.

"Money had to be solicited, buildings had to be erected and the ski hill had to be reconditioned. The largest single contributor and the individual upon whose shoulders rested the final decision to hold the tournament was Mrs. Laura Knight, of Vikingsholm at Emerald Bay and Santa Barbara. Her former husband "Tin Plate Moore," of Rock Island system railroad fame, died and left her substantial means with which to face life's vicissitudes. Her check which ran well into four figures, made the tournament possible.

"It was a tremendous task and the members of the Lake Tahoe Ski Club headed by Norman Mayfield, rendered invaluable services in carrying through the program of construction and organization. Wendell Robie of Auburn was president of the California Ski Association at that time. Members of the Auburn Ski Club, the Truckee Ski club and the Reno Ski Club were indefatigable workers in carrying on the meet. One of the greatest workers in helping make the affair a success was Tim O'Hanrahan, Southern Pacific official, a former manager of the SP hotel in Truckee, who had first agreed to keep it open in the winter of 1909.

"California's Governor Sunny Jim Rolph was present to open the tournament officially and to present many of the event winners with their medals. Anita Page, movie star, was present as the Queen of the Snow Carnival. She was crowned at colorful outdoor snow ceremonies and established on a throne deep in the fir woods overlooking the world's most beautiful mile-high lake.

"As a result of the 1932 National Ski Tournament at Lake Tahoe, winter sports have grown by leaps and bounds."

Winter sports promotion and mountain resort development soon expanded in all directions and, excepting during World War Two, has continued ever since in the snow fields of California and Nevada.

Early sanctioned events included the National Ski Jumping Championship, which delighted a crowd of some 30,000 people at San Francisco's Treasure Island, during the 1940 World's Fair. Following the jump, the Alpine events featured the downhill championship on Red Mountain and the Slalom Championship at nearby Cisco Grove. All events were sponsored by the Auburn Ski club, the National Ski Association of America and the International Ski Federation (FIS).

Olympic Hill became the home of the Lake Tahoe Ski Club, following the 1931 First Annual Divisional Championships and regional tryouts for the Olympic team and the 1932 National Ski Jumping Championships. These were followed by many other tournaments throughout the years.

In advance of the 1960 Winter Olympics the Lake Tahoe Ski club was merged with the Squaw Valley and Reno ski clubs to become the Squaw Valley-Lake Tahoe Ski Club. The club moved its center of activities from the Olympic Hill to Squaw Valley.

At Squaw Valley the expanded club sponsored the 1959 North American Championships to test slopes selected for the 1960 Winter Olympics.

Other early-day organizations were growing in strength. These include the Auburn Ski Club, which in 1931 persuaded members of the California legislature to vote funds for the initial all-year snow removal programs on mountain highways; the Mt. Shasta Snowmen; the Truckee Outing Club; the Placerville Ski Club, the Mt. Lassen Ski Club, the Nevada City-Grass Valley Ski Club, the Viking Ski Club of Los Angeles, and Reno Ski Club.

Competitors for the initial 1930 ski competitions at Olympic Hill came from the old American Ski Association, a short-lived group of paid performers. Many of these exhibition jumpers were to return later to the amateur ranks. Some remained in California and contributed greatly to the development of winter sports.

The first ski meets were limited to Nordic events, which featured only ski jumping and cross-country racing. Alpine events were soon added including slalom and downhill racing, which became popular in 1934.

By the time the 1936 Winter Olympics were staged in Germany, the Sierra was represented on the United State squad by Roy Mikkelsen, Auburn, Clarita Heath, Los Angeles, and ski team manager Dr. Joel Hildebrand of the University of California.

For the 1940 games, called off because of World War Two, the American squad selections would have included Westerners Sig Ulland, Lake Tahoe; Bob Blatt, Palo Alto; Bill Janss, Yosemite; Chris Schwarzenbach, Pasadena; Dick Mitchell, University of Nevada and Hal Codding, Reno and others.

When the international games resumed in 1948 in Switzerland, the invading Americans included Dodie Post from Reno and Bob and John Blatt from Palo Alto.

Making it to Norway in 1952 were Dick Buek from Soda Springs and Dodie Post from Reno. In 1956 the U.S. Olympic Team included Katy Rodolph of Reno and Kenny Lloyd of Bishop.

These great skiers emerged from the high-level competition programs established over the years at Sierra ski resorts. These events were sponsored by the snow-line ski clubs of the Sierra. They featured many annual Nordic and Alpine championships in all classes. Many became famous regional tournaments and races. These included the annual Nordic contests sponsored by the Auburn, Reno and Lake Tahoe ski clubs. Famed shone on the Snow-shoe Thompson Memorial cross-country from Edelweiss on Highway 50 over Echo Summit to the Heavenly Valley ski area.

Names given other major events include the Silver Belt at Sugar Bowl, the Silver Dollar Derby at Reno

California Governor "Sunny" Jim Rolph and Actress Anita Page reigned as King and Queen of the 1932 National Jumping Tournament at Tahoe City. Bert Cassidy, right, Truckee editor and publisher, was tournament chairman.

Ski Bowl, the Inferno at Mt. Lassen, the Far West Kandahar at Yosemite, the Edelweiss Derby on Highway 50 and the Donner Trail Memorial Ski Marathon from Donner Summit to Emigrant Gap.

To encourage inter-club events, the Sacramento Bee of the McClatchy newspaper chain in 1947 established an annual trophy race for teams including two men and two women. This grew into one of the Sierra's most hotly contested races. In 1955 the Bee also launched its annual Silver Ski Race, for junior boys and girls. The races were originally staged at Edelweiss. When that resort closed, the race was moved to Heavenly Valley. this event groomed young racers who later became Olympic team contenders.

Strong youth interest in skiing brought many contests and winter carnivals for collegiate racers. These included events sanctioned by the Pacific Coast Intercollegiate Ski Union.

pioneer snow surveyor

Entry lists of the initial tournaments at Olympic Hill, Truckee and at Cisco Grove, the Auburn Ski Club's development, were well frequented by talent native to the Sierra. Thanks for this talent development go to Dr. James Edward Church of the University of Nevada, Reno, the pioneer of snow surveying. From the ranks of his skiing volunteers who measured the depth and weight of the snow came some of the first champions of the California Ski Association.

On the scientific front, the surveys Dr. Church developed emerged into the California and Nevada Cooperative Snow surveys on hundreds of courses spread across the snow basins of both states.

This work was to become known around the tournament circuit as the Niphometrology Ski Derby. Yes, the Greeks had a word for snow: "nipho".

Truckee jumping team lines up before the meet with mascot, Frank Titus, at far left. From left: Charles Cozzalio, Pete Passinetti, Earl McKay, Jess Maxom and Frank Gaiennie. W.B.B. photo.

The snow study was considered the greatest annual event in the mountains by those who depend on stream flows, such as farmers, ranchers, bankers and hydro-electric officials. Snow surveyors have pointed out that if the potentialities of the western states ever are exploited completely it will be as a result of control of the waters from western streams. They assert every drop, or snowflake, that falls must be held, hoarded and harnessed to make many jobs before allowing it to escape to the sea. That was an idea that simmered in Dr. Church's mind back in the 1890s.

Dr. James Edward Church

Dr. Church, who has become known as the father of snow surveying now enjoys worldwide fame. As a young man he came to Nevada as a professor of the classics. He taught Latin and art appreciation.

He loved the mountains and thus his next role was his interest in the role of snow in its relation to forest conservation.

For Dr. Church it was only a short jump from a hobby of amateur forest conservation to surveying snow depths and forecasting spring runoff for the benefit of Nevada ranchers. In 1900 he strapped on snow-shoes and headed toward the summit of Mt. Rose, a 10,800-foot-high peak 18 miles southwest of Reno. Observers at that time were convinced he was off on a crazy mission.

However, Dr. Church continued his scientific work, piling up statistics. He recruited companions from among his students. Within a few years he had established a complete mountaintop meteorological and snow-testing laboratory. Recognizing the science developed by Dr. Church and the methods used by his original snow surveyors, the world now looks for and depends upon his records and data. These are essential to the wise location and design of the mighty structures that today harness Sierra streams.

Wayne Poulsen begins new ski era

From the Mt. Rose region in 1931 emerged a 16-year-old snow surveyor named Wayne Poulsen from Reno. He was on his way to become a native High Sierra ski great. His competitive career began that winter as a Class C rated jumper in the divisional championships and Olympic tryouts at Olympic Hill. His competitive career was to continue through all the years of sanctioned ski association and intercollegiate events up to and including the 1957 Nevada State Jumping Championships.

During the years in between, Poulsen was to become a ski area developer. In 1939 he launched Reno ski areas at Grass Lake in the Mt.Rose region. Prior to that he had written the original snow survey report, from which came the development of the Sugar Bowl ski area at Donner Summit.

In the post-war years, Poulsen began the development of Squaw Valley. It was in 1935 that he had first visualized Squaw as the site of an American mountain community dedicated to skiing as a way of life.

Where Snow-shoe Thompson was the Pathfinder of the Peaks, Poulsen first tracked skis to survey the snows for science, adventure and resort development. During his undergraduate days at

Wayne Poulsen in 1938 at Johnsville, the same year in which he completed his "Snowsports" survey, which was published by the American Geophysical Union in 1939.

University of Nevada, Reno, Poulsen captained the ski team. He became its star four-way performer. He organized and directed the first Nevada Winter Carnivals and together with Elliott Sawyer of the University of California was in 1936 co-founder of the Pacific Coast Intercollegiate Ski Union. Following graduation in 1937, Poulsen coached the Nevada Wolf Pack ski team for two years. In 1939 the Nevada squad swept all West Coast college tourneys. Its triumph included a

Wilbur Maynard, 1910 manager SP Hotel, Truckee, was the key figure in bringing winter sports to the region. W.B.B. photo.

widely publicized jumping event on Treasure Island during the first week of the Golden Gate International Exposition.

The Nevada team then was credited with being the mythical title of national intercollegiate ski champions.

During tournaments sanctioned by the California Ski Association, Poulsen carried the colors of the Auburn Ski Club. He participated in the many great promotions aimed at selling snow and mountains to the flatland folk. In these events ski jumpers competed on shaved-ice created slides at Berkeley in 1934, the Oakland Auditorium in 1935, the Los Angeles Coliseum in 1938 and during the International Ski Federation Championships which opened the 1939 San Francisco World's Fair on Treasure Island.

Poulsen grabbed center stage in 1938 as the principal figure in two events that won world-wide publicity for the High Sierra.

As the newly crowned Nordic combined ski champion of California and Nevada, a title he won during the divisional tourney at Mt. Lassen, Poulsen led an historic invasion of the Lost Sierra. He was drawn there to pick up a challenge tossed out by surviving members of the Alturas Snow-Shoe Club. Even though armed with imported flying-kilometer and jumping skis, Poulsen and his highly ranked racers, went down to defeat just as Snow-Shoe Thompson had in 1869!

Then Poulsen journeyed to Los Angeles for an appearance before the American Geophysical Union. He read a paper covering a study which classified snow conditions in the Sierra as they would apply to the development of winter sports and resort areas. He was sponsored by Dr. Church, then president of the

E. Des Baillets, Yosemite Winter Sports Chief imported from France.

International Snow and Ice Congress. The ski world has since noted that Poulsen's findings were another great step in the forward march of skiing.

The great ski center that developed, hallmarked by the feats of great snowshoeists of the past, is now known as the Sierra Loop ski center, which includes all the snowfields along U.S. Highways I-80 and 50. It skirts Lake Tahoe and stretches across the eastern border into Nevada, taking in Ski Incline, Mt. Rose and Slide Mountain. It includes Squaw Valley and the Olympic area, Alpine Meadows, near Tahoe City; Northstar-at-Tahoe and Tahoe-Donner, near Truckee; the Sugar Bowl and Donner Ski Ranch, atop Donner Summit; Homewood and Tahoe Ski Bowl on Lake Tahoe; Sierra Ski Ranch above the old Edelweiss area on U.S.50, and Heavenly Valley dominating the south end of Lake Tahoe to be perhaps known some day as Snow-shoe Thompson Memorial Park,

Skiing and other winter activities have made Lake Tahoe, once famed for summer pursuits, an internationally renowned four-season resort.

Yet, La Porte, a forgotten winter wonderland with a hardy population of snow-shoers, rests quietly in the heart of the Lost Sierra, the cradle of California skiing.

References:

Interviews by the author
The Downieville Mountain Messenger
The Virginia City Territorial Enterprise
The Alpine Chronicle
The American Geophysical Union

Olympic champion skier Sigrid Lamming, center, teaches ski joring at Tahoe City, 1932, showing the erect style no longer employed.

19

the Lost Sierra revisited

The Lost Sierra Revisited...the Palmy Days When "Dope Was King and Money Was Plenty" are recalled in answering...The Colonel's Letter.

in search of a story

"Few outsiders ever enter the 'Ski Cradle,' as it is virtually a lost world where life has not changed since gold was discovered on Rabbit Creek in 1850." The La Porte Scrapbook, by Helen Gould, pioneer resident of the Lost Sierra.

In 1938, I ventured into the Lost Sierra and found living history of the Snow-shoe Era. That winter opened with torrential rains, which ended in the most monumental snow depth ever recorded in the Sierra Nevada. The weather turned the calendar back to pioneer days before modern snow plows kept the mountain passes open to a time when isolated Lost Sierra communities depended on sturdy men, equipped with snow-shoes, to bring in the mails, medical supplies and other necessities.

For me there is a kind of nostalgia in reviewing that winter. I was called to action at its very beginning when on Dec. 11, 1937, the heaviest rainstorm of this century drenched the mountains. Floods swept through historic Downieville, washing away numerous homes. Raging water destroyed three bridges, damaged scores of business structures and mines. The loss exceeded half a million dollars.

In the Sacramento Valley, towns were inundated as foaming waters poured down from the mountains. Houses, barns, fences and all kinds of objects were sent whirling along on the bosom of the Yuba, Feather and their tributaries.

For editors of California's metropolitan dailies it was a made-to-order circulation-boosting story. Downieville was cut off from the rest of the world!

Johnny Redstreake, snow-shoe champion, 1938, pictured in 1973. Photo by David Hiser.

Flood disasters have always been historical events.

The earliest information about floods of such terrible impact may be found in the folklore of early day savages. They told the story of thousands of California aborigines losing their lives in the circa 1805 Sacramento Valley flood. Further, the annals of the Hudson Bay Company show that 1818 was a year of excessive storms and raging floods, while during the winter of 1826-27, when Jedediah Smith passed through the northern California region with his trapping party, the water in the Sacramento Valley rose so high that his party was driven to the refuge of the Marysville Buttes. He shared the high ground with elk, antelope and bear which also sought refuge. Adding glamor to the legendary California floods were the more factual accounts of other wet seasons as reported by the early gold-seekers.

Newsmen, that December 1937, already knew that Downieville had been knocked out by flood waters and snow on numerous other occasions and was ripe for the front page.

Soon San Francisco's highly competitive newspapers had reporters flocking to Downieville, all intent on obtaining good stories about the disaster.

By that time the bridges below Downieville had been washed away and all ingress from the west was cut off, at least temporarily. There was, however, a good chance that someone could get in and out by way of the Yuba Pass, approaching Downieville from the east.

a snowbound journalist

A telephone call from the Sacramento Bee soon sent me across the mountains headed west from Reno. I was accompanied by Robert Miller, a University of Nevada student who at that time was a reporter "stringing" for the United Press. He was a fine reporter who later became famous as a foreign correspondent. While our trip into Downieville turned out to be a failure, due to disrupted telephones, it created a situation which led to other events culminating in our discovery of the Lost Sierra and a snow-shoeing renaissance.

The Yuba Pass highway looked like a river as we wheeled down from the Yuba Pass summit to reach Downieville. All the phones were dead. We saw there was plenty of floodwater in town to last until communications were restored. However we noted that the residents were worried over the possibility of snow blocking the Yuba Pass and completely cutting off the town's sole remaining link to the outside world.

By that time our rival journalists were being ferried across the churning Yuba by breaches buoy. Miller and I decided that we were two reporters who were going to get out of there before the rain could turn to snow and block our exit. We took off for safety, fully determined to telephone stories from Sierra City and then proceed across the Yuba Pass to Reno.

Things did not work out that way. In Sierra City we found telephone communication disrupted. Following an hour's wait we proceeded up toward the summit. Unfortunately, the Sierra City delay had cost us any chance of escaping from the mountains that day. A bridge above town had washed out, forcing us to return to Sierra City, where 24 hours later we abandoned our car to hike out across the mountains. At that time my interest in the Lost Sierra snow-shoeing legend had tempted me for several years. I had often regaled Miller with old-timers' anecdotes. During our enforced overnight stay in a ramshackle hotel, a building dating from early times, I had the chance to substantiate my veracity and cement a longtime friendship with Miller, who heretofore had been a skeptical listener.

Among the inn's permanent residents was the town's beautiful school teacher. Several hardrock miners also came down to town from the nearby Sierra Buttes Mine. It was the gently reared teacher's first mountain school experience and she listened in fearful silence to the rain's tattoo, a somber background for vivid tales of avalanches and adventures on snowshoes as related by the men from the mine.

When predictions confronted us with the threat that the rainstorm was a mere harbinger of the greatest snows in history, Miller and I decided that if such should be the case we would emulate, on skis, what the pioneers had done on snow-shoes.

We snapped on our skis and completed the cross-country trek without incident and filed our stories to the waiting editors.

Seven weeks of abnormally warm weather followed our escape. There was little snow piled up anywhere in the Sierra until Feb. 1, 1938, when the season's first major snowstorms rolled in from the Pacific and piled up 7-10 foot depths in 48 hours. The snow fell unabated, hurled by a 60 miles per hour gale. It was not long before all trans-Sierra passes were closed.

Barometers plunged. Storm followed storm until February 12, when the California Highway Department announced that an all-time record snow depth of 227 inches had piled up at Donner Summit. The pack had broken the 1911 record by one inch!

This was "it," Miller and I decided as we swung away from our typewriters.

The morning of February 13 found us in Truckee, where we clamped on our skis. We were determined to ascend to Donner Pass and reach the highway maintenance station during the worst storm in history.

We made it, but were so terrified by the climb that we both decided, thereafter, to leave such perilous travel to others with more pioneer spirit.

Snowslides had roared down from above as we inched upward. Fortunately they crashed either in front or behind us. We struggled over 30 to 50 foot drifts along our steep climb.

On reaching the snowbound highway maintenance station, which Miller dubbed California's "Little America," we found one shift resting while another battled the storm.

For all practical purposes the pass area was isolated. Trans-Sierra Highway 40 was closed and Donner Pass was blockaded. Telephone lines were down at several locations. Trains were stalled and airplanes grounded.

It was a place for strong men. Hardy crews threw themselves into the fight on 12-hour shifts. State employees manned snow-gos, while power company and telephone crews battled blizzards to repair pole lines being blown down by savage winds. Wallowing through towering drifts the workers cursed the state, the companies, the tools and the snow country. Soaking wet or frozen stiff, they groused like soldiers caught in frontline foxholes. There was trouble to spare all during that big blow. The fight continued day and night as smaller storms followed in rapid succession during the next four weeks.

Meanwhile, in the Plumas and Sierra county high country, life was no bed of roses. Mining towns groaned under two dozen feet of snow. This was especially true of La Porte, where by February 28 the food supply was getting monotonous. Worse still the liquor supply for the camp's permanent residents had long since run out. Only first class mail was being delivered weekly thanks to intrepid snow-shoe carriers.

The snowbound populace was in a rotten mood, muttering complaints. What residents wanted was some important second, third and fourth class mail bundles, such as newspapers and a few packages of fresh groceries.

The residents' immediate wants were satisfied March 1 when Warren Shingle, the mail contractor, combined old and new means to get the job done. He snow-shoed into La Porte and ordered residents to lay a red blanket in the town's main streets as a special air mail delivery target for an airplane. Long-delayed reading matter and parcel post floated down, but the snowbound miners were still dissatisfied. They didn't have enough to do.

La Porte townspeople began to dig themselves out in mid-March. They continued shoveling well into April.

snow-shoe renaissance

The time had come for a snow-shoe renaissance.

Volunteer snow-shoe patrols were organized to take the winter's injured out to get medical attention on the western slope.

Meanwhile dopemakers brewed their esoteric Sierra lightning. Squad riders prepared for a race at Jamison Ridge, on the eastern slopes of the mountains.

Hemmed in by deep snow, there arose a stirring sense of excitement among the citizens the like of which had not been seen since 1911. Eager snow-shoers were turning back the clock to the days in the Lost Sierra when dope was king.

Miller and I caught the fever. We soon were headed back toward the Lost Sierra in response to the following invitation:

"Dear Bill, Snow-shoe races at Jamison Ridge on March 20. The gauntlet is down. Bring your skiers." It was signed by Len O'Rourke.

The word spread like wildfire among the snowbound oldtime snow-shoers. Organized modern-day ski racers in outlying valley communities responded to the widely advertised invitation.

Race day brought a goodly field of oldtimers and modern-day ski racers to Jamison Ridge. The site features a 1600-foot race track, topped by a steep starting schuss, and a steep run that then flares out onto a flat outrun.

At the bottom of the ridge stood a crowd of some 500 spectators. The throng was divided equally between oldtime snow-shoe and modern-day ski race fans.

Snow-shoers versus "skiers"

The ski racers, who had been attracted by the advertising broadsides posted throughout valley towns, joked about the 12-foot snow-shoes being exhibited by the oldtimers. They bragged about what their ski champions would show the snow-shoers when the speed trials got under way. They thought the oldtimers' efforts would be a laughing matter when they were pitted against each other on the steep track.

It was old vs new that day. Modern ski racers confidently facing oldtime snow-shoers, shoulder to shoulder.

There were oldtimers who told the invaders that snow-shoers had been king of the trails for many years. They claimed no skis ever made could beat their good old longboard snow-shoes and lightning dopes.

The skiing visitors laughed, pointing to the starting line. There stood Wayne Poulsen, California's newly crowned ski champion, Earl Edmunds, Lenny O'Rourke, a national champion among other ski racing stars. Well-known ski racers included a contingent from the Eagle Ski Club, which, though not so skeptical, had come armed with two pairs of "flying kilometer," racing skis. They were copies of those famed for their use in setting straight downhill speed records in Switzerland.

Soon the air atop the ridge grew tense.

Nearby played out a fascinating picture of secretive dope preparations by the oldtimers. Already observing something mystifying, the modern-day ski champi-

Wayne Poulsen, 1938 combined U.S. ski champion, left, and Earl Edmunds, 1932 U.S. Class C jumping champion, Reno, looked with amazement at towering Lost Sierra snow-shoes at the 1938 revival of longboard racing. W.B.B. photo.

ons stared in awe. Here they thought they saw something new. Maybe the oldtimers knew something they didn't, something tried and true out of the past?

It can't be true, they comforted themselves.

Dopemakers confidently tested a variety of what the champions could see were waxes of some kind. They watched oldtimers feel the snow and sniff the wind. The dopemakers glanced up at the sun, then drew their selection from pouches. They carefully smoothed their strange-smelling concoctions onto their longboard bottoms. A hush fell over the crowd.

Len O'Rourke, a recognized turn-of-the-century champion, called the squad riders together, explaining that racers drew lots and would run four at a time.

Four snow-shoers stepped up to the starting line, preliminary to forerunning the course. O'Rourke informed the entries that they were to follow in squads and saw-off in round- robin fashion to determine the final winner.

As the modern-day ski racers lined up, they nervously voiced their belief that a warm sun might slow the course. Then they watched the oldtimers slide away, tucking their single poles under their right arms and setting their sights on the finish line 1600 feet below.

The forerunners momentarily disappeared from sight as they pushed off from the starting line and dropped down the main schuss. The oldtimers quickly reached the flat. Incredibly, their speed kept increasing as they literally flew toward the finish poles.

"Pretty fast rides," commented the modern-day ski racers. "The snow must be faster than we thought," they murmured.

Then came the rude awakening.

oldtimers' victory

O'Rourke called the first four riders. Two were modern-day racers on their short skis and two were oldtime snow-shoers on their longboards. At the signal the quartet slid away for a good clean start. The four men were all even atop the ridge.

In a flash the two oldtimers shot so far ahead, plunging downward at such a fast pace, that the modern-day boys appeared to be standing still. "Hah," chorused the ski fans, "That was just an accident."

"Those racers were just a couple of amateurs. Wait until Poulsen, Edmunds and the Eagle boys run. They'll show you old snow-shoers something," they chorused.

O'Rourke then called for the second squad to the starting line. It was Poulsen's turn. How the skiers cheered!

Poulsen readied for his ride on heavy eight-and-one-half-foot-long jumping skis. This would be a walk-away all right, the moderns felt. There were only two other skiers and one oldtimer, white-haired, 61-year-old Ab Gould for Poulsen to beat.

The oldtimers stared grimly. Could Gould, who hadn't raced since 1911, possibly summon the skill and daring that had carried him to many victories at La Porte, Gibsonville, Howland Flat and Table Rock?

As the racers lined up, a sparkle lit old Ab's eyes.

He seemed to have been recalling the stories his father told him when he was a youngster. He had been told of how the great Snow-shoe Thompson had been vanquished in 1869 by the Plumas County boys. Thompson had entered the Lost Sierra confident with his towering reputation as a skier, only to find he was slower than the slowest. He was one among many invaders who over the years had never skied fast enough.

There were four racers crouched at the starting gate. Three youngsters pitted against one white-thatched oldtimer. The three held membership as top-seeded racers in the National Ski Association of America. The fourth, whose laurels hung in obscurity among the dwindling roster of the Alturas Snow-shoe Club, gazed down the track. "Here is the ultimate challenge," might have been the thought that crossed Gould's mind as he tensed for a run against the National Ski Association's finest.

Gould must have known that the National Ski Association practically snubbed the snow-shoers way back in 1904. That was at the time when La Porte's Jim Mullen wrote officials in Ishpeming, Michigan, seeking information about entering a forthcoming national jumping tournament. Mullen offered to send a snow-shoe squad east. Further, had not Harold A. Grinden, the USSA's historian and onetime president, finally uncovered that letter and commented on it, it might never have been known that the Lost Sierra was the cradle of Western America skiing. It was in the 1936 U.S. Ski Annual that the letter was revealed as "proof of an early start of skiing in California!"

Few spectators had any idea what was on Gould's mind, or knew anything about how snow-shoers had been ignored in the past. To them the race was going to be a pushover for young Poulsen, who had just returned from an international racing meet in Sun Valley. He was expected to win his first heat easily and rapidly, then advance to the day's final run.

"Atta boy, Wayne!" the other skiers cheered.

old abe hits 70 mph

Gould was grim. He would and could.

The drum was tapped and away they flew.

Ab gave four rapid strokes with his single pole, then assumed a low racing squat.

They were traveling fast as they hit the steep drop. The skiers crouched, gathering speed, but wonder of wonders, Gould's snow-white thatch crept ahead, leaving Poulsen and the other skiers behind. Gould was pulling away like greased lightning. His pace quickened as he hit the flat. The oldtimer became a blur at the finish line. He had traveled 1600 feet in 17.2 seconds. That was a fast time for a course that had a practically level second half.

Gould hit an estimated 70 miles per hour clip as he shot through the winning poles. The oldtimer had, from a standing start, left the modern champion far behind!

So went the rest of the day's squad races.

Not a single rider from the outside skiing world won a heat. They and their downhill, jumping and flying kilometer skis all bowed to the prowess of the oldtimers on 12-foot snow-shoes primed for speed with lightning dope.

Winner of the final round-robin squad was Johnny Redstreake, a 30-year-old mailman from Quincy. He was Gould's protege and had used his sponsor's dope. In the old tradition of the snow-shoe fraternity, Redstreake and Gould, as rider and dopemaker, split the winner's purse.

The crowd cheered wildly. It had been a long, full day of snow-shoeing triumph. Redstreake, who had run the course in 17.1 seconds, remained king!

Stormy weather, which had lifted long enough to stage the races, blew in again. The deepening snow brought much self-pity to townsfolk back at LaPorte. Residents were not happy about a grocery store with no groceries and its two saloons without liquor.

not like the old days

Then came an opportunity to bring their plight to the outside world.

Bud Cosker, a prominent townsman, had been injured in an accident at the Lucky Boy mine. He was strapped to an improvised toboggan. Twenty-four men on snow-shoes formed a powerful line to drag him across 20-miles of snow to safety at Strawberry Valley on Sacramento Valley's western slope.

The La Porters' complaints as to the alleged inadequacy of winter time mail service was then placed in the editorial hands of The Oroville Mercury.

Editor Dan Beebe trumpeted:

"There was a time when La Porte got its mail, snow or no snow. The mailman fitted snow-shoes to his horses and went through the snow, carrying both first class mail and the very necessary parcel post, that included food and medical supplies. Not only were the mail contractors responsible men, but they were paid enough so that they could afford to have good equipment. Since that time, it seems, the bids have been lower and lower. The government always accepts the lowest bid and then has sat by content with inadequate service. We are told that the present mail man has lost thousands of dollars on his contract and that the successful bidder on the contract just let gets the job done for a thousand dollars less. What can be expected as a result? Why, inadequate service, of course. If the payment is too small, let it be made larger, and then let the government see to it that sufficient and adequate equipment is placed on the job to keep the mails moving so that La Porte and other points along the line may be assured of food and mail throughout next winter and from then on."

The article was like a red flag to the editor of the nearby Marysville Appeal-Democrat.

He roared like a bull and claimed to have found a "funny old skeleton in The Mercury's closet. He declaimed that "The Appeal-Democrat will not be a party to any bombast and cant designed to transfer the La Porte mail route to Oroville."

The Marysville editor itemized the situation as follows:

"1. The mail has been delivered to La Porte this winter with an almost uninterrupted regularity in the face of extraordinary natural winter handicaps and hazards.

"2. The present star route contractor, despite restraints put upon him by questionable post office practice of letting the job to the lowest bidder, has available and employs every possible mechanical and human equipment to facilitate delivery, even to include numerous horses on snow-shoes, tractors, airplanes and skiers.

3. We should not be misled by individuals who picture the woes of the permanent winter residents of La Porte, because hardship and privation have not existed there, except for a brief shortage in the liquor supply of certain individuals. To suggest otherwise would be an insult to the intelligence of La Porte's established provisioners.

"4. La Porte, thanks to decades of experience with natural conditions over which we humans have no control, is prepared for winter, takes it when it comes and likes it.

"5. Marysville is far from unwilling to cooperate with Oroville in bettering mail service anywhere. If the Mercury will have its informants point out wherein the present

Art Gould, pioneer snow-shoe mail carrier, son of Lost Sierra pioneer, Jake Gould, is pictured in 1919 on his mail route between La Porte and Nelson Point.

Truman Gould, grandson of Pioneer Jake Gould, shown leaving his La Porte cabin with his snow-shoes in winter of 1972-73. Hiser photo.

service to La Porte is not functioning as to regularity, equipment, human energy and otherwise insofar as insurmountable conditions permit, Marysville will take the whole job off The Mercury's hands and command correction."

The Marysville editor had taken a firm grip and would not let go.
His headlines invoked:
NO DOUBT MR. FARLEY WILL STOP THE SNOW AND LET THE MAIL GO THROUGH
Oroville Mercury Suggests Oroville and Marysville Take Steps to Relieve La Porte.

"Mr. Farley, one of the best little fixers we've ever had as head of the postal service, will be quite willing to arrange for an Oroville dairy to deliver whipping cream to La Porte doorsteps each morning, and, of course, to supply the already well known La Porte larders with French pastries for each gathering of the Tuesday afternoon club."

Caught in the midst of such a warm battle, the emissaries from La Porte soon snow-shoed back to their isolated homes to sit out the snows until the coming of summer.

Not until mid-July, 1938, was the road from Johnsville to La Porte thawed out sufficiently to permit trans-Sierra travel.

It was then that I set out for La Porte in search of James F. Mullen, who in January, 1904, had written to the National Ski Association, regarding a ski jumping competition.

the colonel's letter

On the steps of the Union Hotel, undoubtedly the world's oldest ski hostel, I found four oldtimers. Jim Mullen was one of them. He was affectionately called "Colonel Bull."

"Colonel," I inquired, "did you ever write a letter East about sending a squad of jumpers to Ishpeming?"

The "Colonel" looked me over carefully and said: "Holy mackerel, why do you ask that?"

"Well," I replied, "If you did, and if you never got an answer, maybe I've got an answer now."

The "Colonel," turning to his companions,
said: "Remember that letter we sent to Ishpeming and how we cussed the Norwegians for not answering it? Well, this young fellow says he's got an answer now."

"Colonel," I began to explain, "I actually haven't got the answer for you. I just know that you wrote that letter. What I want to find out is just what kind of a ski team did you want to send East?"

I pulled out my 1936 American Ski Annual and turned to Harold A. Grinden's article titled: "Away Back Yonder," in which the historian had credited the Aurora Ski Club of Red Wing, Minnesota, organized in 1886 by a group of Norwegians, as being American's first ski club. Grinden had referred to Mullen's letter as he wrote: "This is proof of early California (skiing) activity."

The "Colonel" took one look, and then roared: "If they had only answered that letter then, we could have showed the world!"

"Early activity, eh?" he snorted, "Why when I wrote that letter snow-shoeing was in its second childhood

Col. James F. Mullen

up here. I wanted to head a squad of jumpers and go East. We weren't sissies up here, and I wrote to find out what the financial inducements were. We wanted some traveling expenses paid. None of us had ever jumped in competition as all our events were down-mountain speed races, but we sure knew how to jump, too. Of course, we're all old men now, and there aren't any youngsters coming up, but in those days we still had enough men to have taken on all the Norwegian boys back in Ishpeming."

Mullen and the other oldtimers looked me over.

"You call it skiing or snow-shoeing?" they challenged me.

I looked the Colonel in the eye and replied: "I know how it is. I call it snow-shoeing."

"Come inside then," Mullen invited. "You must be on our side. We'll tell you all about it."

Inside the bar they sat me down and began to spin snow-shoeing yarns from the very earliest gold rush times down to that very day.

Years back they went and names and dates rattled from their memories. They told tales of Washington Hill, Onion Valley, Johnsville, Eureka Peak, Jamison Ridge, Snowshoe Flat, Bunker Hill, Howland Flat, Table Rock, Port Wine, Gibsonville and La Porte. There were real snow-shoe tracks following the days of old and the days of gold in 1849. Out of those legends I have pieced together several historical articles. The first of these was entitled "Into the Ski Cradle." The nationally published piece resulted in a memorable letter being delivered to "Colonel Bull," the following winter by a snow-shoe mail carrier.

Written on National Ski Association of America stationery and dated December 29, 1938, it read:

"Colonel" James F. Mullen
La Porte,
Plumas County, California
My Dear Colonel:

I have just finished reading the fine Sports Illustrated article, "Into the Ski Cradle," by Bill Berry. I now hasten immediately to send this message to you.

Little did I realize what would happen after my recording the mention of your having written to Ishpeming for ski news away back in 1904. I found this little mention in a local paper and added it to my recordings, part of which were presented in our ski annual of 1936. Then along comes Mr. Bill Berry, who loves to dig up ski history, to follow through with the great story of your section.

Had I been secretary of the Ishpeming Club back in 1904 I certainly would not have let your letter go unanswered. To have mention of your letter come back to life 34 years later through Bill Berry's story is truly one for Mr. Ripley's "Believe it or Not!" classics. If this letter may act as an apology for the early

C. James F. Mullen, center, La Porte, 1892 snow-shoe champion, whose letter published in the January 18,1905 issue of the Marquette, Michigan, Mining Journal, tipped off U.S. Ski Association Historian Harold A. Grinden, that U.S. organized skiing dated back to 1867, not 1882, as ski historians had previously concluded. Grinden thus learned, the historic "first" goes back to 15 years earlier, when in La Porte, heart of the Lost Sierra, the Alturas Snow-shoe Club staged the first organized ski race in the U.S.

From left, three former La Porte snow-shoe champions; Henry W. (pop) Hewistt, Mullen, Mike Bustillos. W.B.B. photo.

club secretary at Ishpeming, for not having answered your letter, then will you accept my apologies, for in a way, as historian of the National Ski Association, I have somewhat taken it upon myself to preserve the work of the past and should accept, I presume, the slip-ups made by some of the pioneers.

Each year has brought on new angles and we have learned much and now that I have just completely read Bill Berry's article, I am hungry to hear more of the early days in California. By piecing together bits from here and there we shall soon have the authentic history of American skiing, and right now it looks to me like California will have a firm place in my official American Early Ski Recordings.

Up until today the only mention I had of California in the early days was the story of Snow-shoe Thompson's mail carrying.

The Ski Club of Berlin, N.H. is listed as America's earliest organized ski club and this occurred in 1882, on January 15, to be exact. Then on January 19, 1886, the Aurora Ski Club, at Red Wing, Minn, was organized. This club is still functioning as is the Ishpeming group formed in 1887. Other early clubs included the Den Norske Turn og Skiforening of Minneapolis, Minn., and a group in Altoona, Pa., both organized in 1885, a year in which skiers also were in evidence around St. Paul, Minn.

I mention these few dates because I am leading up to something that will be of great value to the Historical Department of the National Ski Association.

That lies directly in you and what you might be willing to add to our archives. If only we were close enough and you were willing, so that we could both sit down for a time and check back to those early days. Authentic facts must come from men like yourself and we want so badly to preserve the work of the pioneers. Stories of the olden days, authentic dates, photos, etc., are of vast importance to those historic days. The true stories of Plumas County, La Porte, the old

Harold A. Grinden, Duluth, Minn. lifetime U.S. Ski Association Historian, and founder of the National Ski Hall of Fame. He found the 1905 newspaper notice of Col. Mullen's letter.

Union House Hotel, La Porte, July 28, 1989 photo. The place where the author met Col. Mullen and his fellow snows-hoers.

hotels, the names of old skiers, and their competitions and above all the dates will help so much.

Bill Berry's article has started me on a new hunt. It now looks very much like a new chapter is to be written, and a chapter which will head up what already is preserved as far as I have been able to go. Some day a great American Ski Historical Volume will be published and we must have it authentic and must by all means have a true picture. California and your section, I am sure, will head up that volume.

I have traced skiing from the fifth century in countries across the sea. We now have fairly complete records since 1885 in America's mid west, but those earlier days in California must yet come further to light.

In reading up on early discoveries I checked back to find the story of Leif, the son of Eric the Red, who visited American shores in the year 1000 and also found where Thorfinn Karlsefni founded a colony on the Atlantic Coast about the same time. This led me to say a year ago that skis first reached our shores when these colonists settled. Principally because I felt, and still do, that these early Vikings could not have lived through the winters and heavy snows without the use of skis for travel. In the first part of my letter I apologized for the secretary at Ishpeming. As I finish this letter I am tickled pink that your letter was never answered.

You missed a great trip East because of that letter, but that miss has added much to ski history. Thanks to you, and yes, to Bill Berry...

Yours very sincerely,

Harold A. Grinden

U.S. Ski Association Historian. Harold A. Grinden, unearthed this January 18, 1905 Marquette, Michigan Daily Mining Journal, which reported the Ishpeming Ski Club's receipt of Col. James F. Mullen's letter. Grinden wrote about the letter in the 1936 American Ski Annual. It caught the author's eye.

EXCERPTED FROM

THE DAILY MINING JOURNAL
January 18, 1905

Ishpeming Department

GREAT SKIERS COMING

America's Best Riders Will Participate in the Tournament.

The officers of the Ishpeming Ski club are already assured that practically all of America's best ski riders will participate in the tournament here on Washington's birthday. For the past few weeks the officers have been looking up the records of experts who have signified their intention of contesting. All of the men who took part in last year's meet; including the Red Wing riders are coming and a number of others who have excellent records are included in the list. From present indications there will be from twenty to twenty-five riders, counting the best local men, in the contest.

A.H. Axtell, a merchant of Flatrock WI., writes that his brother-in-law, a young man recently from Norway, is with him and that he will be here. He has taken a number of first and second prizes in Norway. His best jump is given as 104 feet.

Ole and John Maugrett, who won fame in the Norwegian ski tournaments a few years ago, and are known by reputation to every son of the Scandinavian countries who follow the ski contests, will be here. They now reside at Fredericks Wis. One of the brothers is said to have a record of 120 feet, made two years ago in Norway. J.C. Johnson, another celebrity from Norway, now living in a little town in Minnesota has been heard from, and the club has decided to have him attend. Carly Jacobson... a Norwegian with a good jumping record... will be among the contestants. The Kanabas man has a record of over 100 feet for a standing jump.

...There has been much discussion among skiers since Monday regarding the new hill...skiers think the hump is not properly sloped...

<u>James F. Mullen, a California skier, has written the president, stating that he would come to Ishpeming with several riders if expenses are paid...The expense of bringing the California riders here would be too great...the list will be large enough without them.</u>

It was in 1954 that surviving members of the Alturas Snow-shoe Club had come to Reno to watch a new generation of national ski racers. They cheered the young racers. Then the oldtimers sang their own swan song.

"hello snow-shoer! how's your dope?"

At daybreak Dave Hall, a 45-year-old snow-shoe mailman, poled out of Strawberry Valley. He was determined to make the snow hiss like tearing silk beneath his long spruce boards all the way to La Porte. For Hall it was a dual role. Like his pioneer contemporaries who found that snow-shoeing could be fun as well as work, he had just combined the two. Not only was he carrying the mail, but he was also the club's new vice-president.

It was a blustery day and snow was being funneled down from the Sierra directly into Hall's face as he ascended the pine tree-lined route. This was the same trail used by Hamilton Ward and James Murray for their legendary escape on barrel staves back at La Porte in 1850. Over the snows Hall poled into the long ago. His was a centennial snow-shoe trip in California's centennial year.

It is one that is overlooked by those paid to create the hurrah for the state's tourist industry. California's Centennial Celebration was in preparation two years. There were pageants in the Hollywood manner and special newspaper editions rich in advertising and memorabilia. But, strange as it seems, there were no portrayals or newspaper reports telling of the many of the Argonauts who came to dig for gold and remained to create a way of life in the silvery snows.

Being by-passed twice in one hundred years had been just too much indignity for the snow-shoers. First the romantic writers passed them up, then the Centennial Celebration planners did likewise. So the Alturas Snow-shoe Club was reactivated. Its message was carried by Hall up into the Lost Sierra instead of down to the flatlands below.

The word carried by Hall was that the club planned to restore La Porte as a living monument to the legendary days of snow-shoeing in the early mining camps. It was hoped that through promotional efforts contemporary skiers could relive the days when dope was king and money plenty.

Hall shouted greetings to isolated cabins along his way. The time-honored cry of the mountain country echoed back through time: "Hello, snow-shoer! How's your dope?"

Hall halted momentarily at American House to chat with Mrs. Elizabeth Merian, the club's new secretary and historian, justly famous herself as a snow-shoeist active throughout the region. Mrs. Merian was a third generation Lost Sierran. Her grandparents were pioneers of Whiskey Diggings and Poker Flat. She is a native of Onion Valley. For 25 years she carried the mail during the snow-free months to those places. She had often used snow-shoes to assist Hall in the winter time. She could and did tell mailman Hall about snow conditions, up above before he plunged onward toward La Porte.

Hall well knew that none of the early-day snow-shoe messengers endured the hardships of mountain travel only for the pay they drew. As he ascended Lexington Hill, then glided into La Porte, he was pleasantly conscious that something new has been added to his daily work.

Not for money alone did Snow-shoe Thompson carry the mail from Placerville to Genoa. Neither did Lenny O'Rourke, Art Gould, Pap Swigart, Jim Larison, Davy Berry, Zachariah Granville, Creed Haymond, Robert Francis and the many others who once traveled across the snows of the Lost Sierra. They regarded themselves as professional sportsmen and were so esteemed. They wore halos of well-earned glory and were loved for it. Hall knew he had joined the fraternity when he arrived in old La Porte.

the next century

By the end of 1949, the Snow-shoe Era rounded out to a full century. The revived Alturas Snow-shoe Club hoped to resume its once famous place in the snow sports world inspired by a history of thousands of snow-shoers inscribed for posterity in the ancient burying places at Hepsidam, Poker Flat, Howland Flat, Saw Pit Flat, La Porte and other remote Gold Rush communities. Above Rabbit Creek on Waite Hill the roster reads: Steward, Hillman, Ah Pock, Primeau, Dubuque, Cosker, Cayot, O'Rourke, Williams, Wentworth, Berry, McLaughlin,

Jones, Kelly, Mighel, Scott, Kleckner, Hendel, Pike, Mason and other family names famed in the snow-shoers' eternal constellation.

The plans became full blown in March of 1941 when the Historical Committee of the National Ski Association sponsored a return meeting on the Sugar Bowl slopes in the Donner Summit Region.

The second speed races proved to be a repeat performance by Snow-shoer Johnny Redstreake, who showed the heels of his long spruce boards to Hannes Schroll, former national downhill ski champion.

Snow-shoers reactivated the club in mid-December of that year. The permanent reorganization was affected in Brownsville, just below the snowline, on Sunday, Jan. 9, 1950.

sierra nevada ski centennial

On March 27-28, 1954, the Alturas Snow-shoe Club joined the Reno Ski Club to celebrate the Sierra Nevada Ski Centennial.

The Snow-shoe Era club staged the Reno ceremony in conjunction with the Golden Jubilee Year Giant Slalom Championships of the National Ski Association of America.

Participants included National Ski Association President Albert Sigal of Yosemite, surviving members of the Alturas Snow-shoe Club and young competitors entered in the National Giant Slalom Championships staged that weekend on nearby Slide Mountain.

The centennial celebration turned back the clock 100 years to 1854 for Alturas Snow-shoe club members, when snow-shoes, as skis were called in those days, were introduced in La Porte, hub of the Lost Sierra.

The following is an excerpt from a story by Bill Berry, on the 1854-1954 Sierra Nevada Ski Centennial which appeared in in the 1955 issue of the American Ski Annual:

"The race and pioneer celebration are over, but the memory lingers on in Nevada for the National Ski Association's Golden Jubilee Year Giant Slalom Championships and the Sierra Nevada Ski Centennial Celebration staged by the Reno Ski Club, March 27-28, 1954.

"Pioneers of American Skiing—the veteran members of the Alturas Snow-shoe Club of La Porte, Calif.—joined forces with representatives of the National Ski Association, the Far West Ski Association and the sponsoring Reno Ski Club to line a race course in the spectacular Reno Ski Bowl. The race start was heralded by a dynamite blast, as Dean Perkins and Jerryann Devlin, amateurs representing the Sun Valley (Idaho) Ski club, sped to victory.

"They were crowned 1954 Giant Slalom Champions of America out in the Mountain West where such events began many years ago.

"The evening's cheers rocked the Mapes hotel banquet room as Perkins, a Sun Valley ski patrolman, Ogden, Utah, and Miss Devlin, personable young lass, Lake Placid, N.Y., made three trips each to the trophy presentation rostrum.

"First they were awarded Gold Medals of the National Ski Association; next came championship belts. The silver belt buckles on which were implanted the State Seal of Nevada in solid Gold—and then certificates of membership and silk racing colors and stars, of what is probably the world's very oldest and most active ski club, the Alturas Snow-shoe club of La Porte.

"On hand for the ceremony was at least one skier who had been born before the Alturas Club placed skiing on a formal and organized basis for the first time in 1867. That was Andy Swingle, 91, Quincy, Calif. Making the final award presentation was Neil "Buck" Mullen, born to skis 85 years ago, a junior champion in 1885 and champion of the world at downhill in 1893.

"No doubt, Perkins and Miss Devlin thrilled to the honors, won by them in as spectacular a race as any staged in the Sierra Nevada's 100-year-old history of the sport.

"Some 200 fans, including a majority of the remaining oldtimers of the Alturas Club, had lined the slide Mountain race course, where Tourney Director Hal Codding set a course dropping through the heart of Reno Ski Bowl—59 gates in a vertical drop of 2,300 feet—and encompassed by Perkins in a clocking of 2:05.1. Next best of the 62 men starters were Dennis Osborn, junior flash from June Lake, Calif., 2:11.0. while Jack Reddish, Olympian and pre-race favorite finished third in 2:11.2

"Down the shorter women's route, Miss Devlin flashed to victory in 1:39.0. She was closely followed by Dorothy Modenese, Seattle, Washington, 1:40.4 Taking third was Sally Neidlinger, a former Olympian, now residing in Los Angeles, with 1:43.4.

"But the race banquet cheers were as nothing compared to the opening salute Saturday night when the Reno Ski Club turned back the pages of history with a Sierra Ski Celebration buffet dinner in the Redwood Room of Hotel Riverside.

"Appearing for the Centennial Celebrations were the last remnants of not only the Alturas Snow-shoe Club, billed as the world's very first formal ski association, but such other pioneer organizations as the Poker Flat Snow-shoe Club, Table Rock Snow-shoe Club, Gibsonville Snow-shoe Club, Yuba Gap Snow-shoe Club, Plumas Snow-shoe Club, Johnsville Snow-shoe club, Silver Mountain Snow-shoe Club, Port Wine Snow-shoe Club, Whiskey Diggings Snow-shoe Club, Onion Valley Snow-shoe Club; and other contemporaneous groups—all of which sprang up informally during the eighteen fifties and early sixties to form the nucleus around which the formal organization of the Alturas Club was accomplished in 1867.

"The veterans of early ski days were presented as they stood on a rostum back-dropped by blue drapes and a golden eagle from the Alturas club house in La Porte. All this flanked by the flags of membership nations in the International Ski Federation (FIS). On display were 12-foot spruce racing skis and secret 'dopes' which have made the old-timers unbeatable for many years.

"Banquet room rafters raised as Will Primeau, who 50 years ago, while wearing the Alturas Snow-shoe Club colors, won the world downhill racing championship' on Lexington Hill, La Porte, was persuaded to display the belt he was awarded that historic day.

"The veteran of an era when speed was speed and winners had to better 100 feet per second for the overall distance, explained how the shoes were cambered and grooved, how the 'dope' was mixed and cooked. He told why the stars on the championship belts have no points on them—and many other fine details of the early days that had all but been forgotten—in the 50th Jubilee Year of the National Ski Association.

"The ski gang was breathless as it heard how the custom in pioneer times was for the champion skier to receive his fittingly engraved belt award at a party the night of the race—a tradition instituted officially in 1867.

"The champion, of course, was obligated to dance with all the eligible young ladies at the party—and it was found that the pointed stars scratched the girls, or tore the lace on their dresses. So, as a chivalrous gesture, the star points were rounded in 1868 and have been so each year ever since.

"Also on hand was Neil (Buck) Mullen, 85, born to ski in La Porte, veteran of junior racing in 1885, and credited with having been champion of "the whole wide world," during the speed season of 1893.

"The twinkling little Irishman, dean of the world' downhill racers, told his auditors that the youth of today was pretty good when it came to stylish riding, but he recalled, with relish, how modern ski riders had failed to defeat old-timers during the old-fashion race revivals in 1938, 1941, 1951 and 1952.

"The vast audience of skiers came to its feet, as the group of the America's oldest skiers joined in a march to the rostum.

"Leading the procession was Mrs. A.T. (Elizabeth) Merian of American House, secretary of the Alturas Club and herself famed for skiing the mail to Onion Valley, Saw Pit Flat and Poker Flat, and her husband, A.T. Merian, the club's 'dope' expert.

"Then came Andy Swingle, 91, and his 92-pound wife, Birdie. Andy is a veteran stage coach whip and oldest driver who knows the secrets of guiding snow-shoe equipped equines through the fleecy white of the Sierra. And Birdie Swingle recalled how she and Andy rode tandem on fourteen-foot-long snow-shoes in pioneer days—not to mention that she was born in a mountainside log cabin in the long ago.

"On hand, from Norway, with his eyes bugging, was Jakob Vaage. He is the curator of the Norwegian ski museum, Oslo. Jakob, however, declined to bow down to the Alturans claims despite photostatic proof displayed all about the banquet hall. He was busy, nevertheless, measuring the long boards of the Sierra and sniffing the secret dopes in an effort to solve their mystery.

"Skis—or snow-shoes, as they have always been called in the early American West—were introduced to the Sierra Nevada at some date after the winter of 1853-54. How some unsung pioneer created them or brought them in, and then how the American miners improved upon them, has been told by this reporter, as accurately as possible, in an earlier issue of this publication.

"Racing is said to have begun in 1855. Next came informal clubs, many men claiming to be champions, featuring boisterous conduct and gun-toting rivalry.

"This was resolved in 1867 by the founding of the Alturas Snow-shoe Club, which in February of that year, staged the first official downhill speed races for the championship of all snow-shoers to place the sport, for the first time, on a formally organized basis.

"To tie together, the National Golden Jubilee and the Sierra Ski Centennial, the Reno Ski Club called upon and received the backing of many officials of the Far West Ski Association, the National ski Association, the Reno chamber of Commerce, the Plumas County Chamber of Commerce, the California State Chamber of Commerce Winter Sports Committee and numerous participating Sierra Nevada ski clubs.

"Two individuals, however, came forward at the last moment to assure success of the old-timer events. They were Mert Wertheimer, operator of the Riverside Hotel theater restaurant and Fred Shield, the hotel's publicity director.

"Wertheimer placed all of the personnel of his vast organization at the disposal of the Reno Ski Club and there was a reason for his benefaction. His birthplace is close to Ishpeming, Mich., where the NSA was founded 51 years ago. so it was only natural when Wertheimer heard of the NSA Golden Jubilee and Giant Slalom Championships that he should invite his friends among the Sierra Nevada oldtimers to join hands in Reno with the ski gang to honor the Alturans—the snow-shoers who back in 1904 wrote to Ishpeming for information on a ski jump meet and then bobbed up again five years ago as the parent body of all skidom and bestowed honorary membership upon the NSA, the FWSA and FIS."

longboard revivals

In 1905, the National Ski Association was organized at Ishpeming, Michigan. To perpetuate this history in the annals of the Golden State and the nation, snow-shoe veterans and supporters staged several race revivals organized by the NSA's Historical Committee. These included four snow-shoe race revivals, all of which were staged with the approval of Roger Langley, NSA President, 1937-48, NSA secretary, 1949-55 and for many years editor of the American Ski Annual.

The four revivals were widely reported by the author for many publications including, the Nevada State Journal, Sacramento Bee and the American Ski annual.

The first snow-shoe race revival was celebrated at Snow-shoe Flat, Johnsville, 1938. The classic contest between oldtime snow-shoers and modern ski racers was widely pictured and reported. It featured national champions such as Wayne Poulsen and Earl Edmunds. The meet was chaired by oldtime Snow-shoer Lenny O'Rourke and won by Johnny Redstreake.

The second snow-shoe race festival was staged at Sugar Bowl in 1941, which was again won by Redstreake. Again the race pitted descendants of the Lost Sierra Snow-shoers against modern champions. In winning the race Redstreake outpaced the Austrian National Downhill Champion and U.S. Downhill Champion, Hannes Schroll.

The third longboard extravaganza brought wide attention to the Lost Sierra capital of La Porte in 1951. This downhill race on Lexington Hill saw the defeat of legendary members of the Auburn Ski Club; Roy Mikkelsen, Ron Mangseth and Orrin Ellingson, all former ski champions. They watched longboard rider John Cowley, University of Nevada, Reno, student schuss off with the prize money.

The fourth longboard competition's return April 5, 1952 brought champions to Snow-shoe Flat, Johnsville, for the "86th Annual World Championship for the mountain country," and delighted a huge crowd. Jerry Burelle, son of Snow-shoe Era Pioneer Medrick (Jerry) Burelle, won the coveted championship belt.

billy kidd

Another revival received wide attention because it featured U.S. Ski Team Olympic Medalist Billy Kidd.

The widely publicized Snow-shoe Era finale came in 1964, during the Nevada Centennial celebration of its statehood. The event, chaired by Reno's Don (Snoshu) Thompson, featured Kidd, who was defeated in the match race by Jerry Burelle of the Alturas Snow-shoe Club. The event, staged at Tahoe Meadows, alongside the Mt. Rose Road, was strictly a wax race, where Burelle a descendant of a Snow-shoe Era pioneer, again proved that "dope was king."

Contestants madly poled off at the sound of the drum in a simultaneous start. Officials had agreed in advance that there would be no skating. Burelle's fast dope propelled him to victory.

It was the last hurrah for "Snow-shoe Era" racing, which had started with the first organized race staged by the Alturas Snow-shoe Club at La Porte in 1867.

In December, 1989, Don Thompson, better known as Ski Writer "Snoshu" Thompson, recalled the 1964 race:

"I can still see the look on Billy Kidd's face when he lost the downhill race to Jerry Burelle as part of Nevada's Centennial Celebration. I was in charge of the race that pitted Billy against the 12-foot boards under Bruelle, representing the last efforts of the Alturas Snow-shoe Club to keep its history alive. It did, in 1964.

"I still see Billy at various celebrity ski functions. He still doesn't believe he lost. Someday we'll have to show a new generations of racers how to really wax their skis."

But it was the emotional grand finale for surviving members of the famed Alturas Snow-shoe Club that will always be remembered. The Pioneer snow-shoers were feted during the same weekend as the March 27-28, 1954, National Ski Association's Golden Anniversary, Reno.

It was on that date that grizzled oldtimers of the Alturas Snow-shoe Club had come to Reno to watch a new generation of national ski racers. They cheered the young racers, then sang their swan song.

final rememberances

America's entry into World War II halted reorganization of the Alturas club—however a fall 1949 visit to La Porte and Poverty Hll by the Irving Berlin family played a large part in creating a new generation of interest in the racing hub.

The highly publicized visit happened because Irving Berlin's wife, the former Ellin Mackay was researching for her novel, "The Silver Platter," based upon the life of her grandmother, born Marie Hungerford.

. She was particularly interested in the legendary snow-shoe tracks of her fabulous grandparents, "Bonanza" John and Marie Hungerford Mackay. Mrs Berlin had learned that her mother in 1868 had raced across spring snowdrifts to the Lost Sierra by pack train in a vain effort to save the life of her critically ill first husband, Dr. Edmund Bryant.

Mrs. Berlin persuaded her daughter, Mary Ellin, to come along. I was their driver and unofficial guide.

Until we arrived, no Mackay had visited the Lost Sierra since Bonanza John had left Sierra County's Forest City and snow-shoed eastward over Henness Pass and the Nevada mountains to Virginia City.

We left Reno: Ellin Mackay Berlin, Mary Ellin Berlin, my wife Frances and I. We took Highway 395 north to Hallelujah Junction, then turned west on US70 over Beckwourth Pass to Blairsden, then entered the snow-shoe country at Mohawk. We rolled up past

Johnsville, a Snow-shoe Era landmark squatting below towering Eureka Peak.

We climbed the thickly wooded mountainside. The one-way, low-gear road wound alongside a precipitous canyon and forded occasional streams that rushed across the rubble-strewn roadbed. The forest primeval revealed only an occasional scar, evidenced by a few mine dumps and an occasional crumbling cabin. There was little trace of the thousands who once scratched for gold among the rocks and creeks. The trail had reverted to primitive dirt, as it was when the emigrants labored that way with their wagons in 1851.

When we reached Whiskey Diggings the trip turned into a motor cavalcade. Word had gotten around among the old-timers that the Mackays were coming back to La Porte and Poverty Hill. We were greeted by Mrs. Marguerite Delahunty, the septuagenarian "balsam girl" of Gibsonville; Mrs. Elizabeth Merian, the Poker Flat mail carrier, and others of the snow-shoeing fraternity. In a cloud of dust we rolled onward through Gibsonville to La Porte, the heart of the Lost Sierra.

At La Porte, a gold rush town originally called Rabbit Creek, Mrs. Berlin again picked up the trail of her grandmother, a trail that she had been following through records and legends for many months. It was in the La Porte cemetery that Marie Hungerford had in 1866 buried her first husband, young Dr. Edmund G. Bryant.

Marie bore Mackay a son, Clarence, who was destined to become Ellin's father. Ironically he later tried vainly to block her marriage to Irving Berlin, the cantor's son. That was in 1926, as Ellin well remembered. Too, as she stood on the site of once-famous Poverty Hill, where Dr. Bryant had been succored by Marie, Ellin recalled some of the tales her grandmother had related before she died in 1928 at 85. Although Mrs. Berlin now has many notes on her grandmother's life and times in the mining camps, she said the old lady preferred to dwell on such matters as her friendship with Queen Alexandra and Edward VII. She had become a great lady and a celebrated hostess in the years following the day when Creed Haymond passed the hat among the snow-shoers of La Porte to bury Dr. Bryant.

On the site of the cabin where Dr. Bryant was found by Marie in Poverty Hill looms a huge mound of white quartz debris. It is a pile which gold snipers, working nearby, said was valued at 50 cents a cubic yard. We climbed to the top and looked out over the old abandoned hydraulic diggings.

Ellin picked up a handful of the debris and let it dribble though her fingers. She smiled at Mrs. Delahunty and said:

"My grandmother never came back to these mountains, but I'm here now..."

Leonard Burr O'Rourke

No account of the Alturas Snow-shoe Club's reactivation would be complete without additional details about the life of Lenny O'Rourke. Leonard Burr O'Rourke was an old and dear friend of the author. It was Lenny who had always believed that some day the Alturas Snow-shoe Club would be

Ellin Berlin, left, and her mother Ellin Mackay Berlin, shown at Poverty Hill standing on a mound of white quartz debris. They study the cabin site where Hungerford, Mrs Berlin's grandmother, found her first husband, Dr. Edmund Bryant as he lay dying.

resurrected in all its former glory, but since June of 1949 he had been sleeping in Whispering Pines in the eastern foothill country. He was the father of the modern Feather River Highway and many other good public works in Plumas County. He had served the county for many years in various official capacities.

It was fitting that during Lenny's Masonic services the last of the previous winter's snowpack could be seen shining on the Sierra's crest, for Lenny had been of that breed of men who fostered the birth of American Skiing in the Lost Sierra during the Gold Rush days. Skiing was cradled in the hearts of these pioneers as an inherent part of their own lives. It became their way of life.

Lenny was no giant in stature, but, nevertheless, he was a heroic figure in the skiing world of the late nineteenth and first half of the twentieth centuries. He was a native of La Porte, born in 1878.

He could command snow-shoes almost as soon as he could walk. At 16 years of age he was packing supplies and mails to the isolated mines and camps. Soon his ability caught the eye and fancy of Frank Steward, the old master dopist of the Alturas club. Under the coaching of Steward he became one of the town's top-ranked racing stars. The association with Steward continued until the old man's passing in 1913.

It was Lenny who in 1938 tossed the gauntlet to skiers, provided the dopes and then stood by them while the snow-shoers sped to victory. Further, he led the snow-shoers on their successful invasion of the Sugar Bowl three years later. He was the possessor of Steward's secret dope recipes and knew how to manufacture them. Lenny's last snow-shoe trip was to Spanish Peak and the Buck Island Lake country in late April of 1948 when he took the author along for some dope testing.

Lenny was a great believer in organized skiing. He attended several championship events of the National Ski Association during the last 20 years of his life. Often he tried to interest ski officials in the snow-shoers' history, but mostly they thought he was just "yarning." He and the author reminisced about those setbacks during that final snow-shoe trip. Lenny then talked about the great winds that blew during the Snow-Shoe Thompson controversy.

"Despite the fact that the La Porte boys took Snow-Shoe Thompson into camp," he said, "we of Plumas County always had plenty of regard for his grit and courage. Who should know better than we of the difficulties he faced in carrying the mails? Thompson has his place in the history of skiing in the United States. We cradled the sport and he obtained the first national recognition, which we believe he deserved. Our center was so localized, isolated and inaccessible that when mining operations slowed down and the organized races were abandoned, well, everyone just forgot about our ski racing center. I guess the fact that we called everything snow-shoeing worked against us. We had our day. I think we will have another."

Charles W. Hendel

Charles W. Hendel was another giant of the Snowshoe Era. Hendel was the well-known "Quicksilver," whose Snow-Shoe Era 1874 stake racing reports in The Downieville Mountain Messenger and his illustrated article in the 1874 issue of Scientific Press, San Francisco, have greatly contributed to this book. (See Chapter 4).

The week previous to his death he made plans for building a concrete cellar and for placing a slate roof on his dwelling. He also contemplated to purchase a typewriter and an adding machine. He was a looking forward, not backward, optimistic, practical and hopeful, a truly remarkable man for his years. His eulogy, given September 17, 1920, reflected this:

"In business dealings, he was honest and considerate. As a friend he was loyal and helpful. With a cheerful greeting and a hearty handshake for all, he counted everyone his friend and harbored no petty grievance, but tactfully averted any indication of displeasure by some timely personal experience or comparison. He delighted to relate reminiscences and his memory was wonderfully clear and accurate,. His stories were told in such a forceful and highly interesting way that he never became tiresome even though he repeated the story. In his long and strenuous life he had many narrow escapes from death and serious injury. History states that he had a miraculous escape from death when he plunged headlong to the bottom of a mine shaft, fifty-four feet deep, his injuries confining him to his bed for only two weeks.

"His wonderful endurance was his greatest asset, for no one ever attempted more hazardous journeys over the mountains on snow-shoes than did he. Even during the past 10 years he prided himself on being one of the few Supervisors of the State that covered his territory on snow-shoes. Chas. W. Hendel will be greatly missed by his colleagues when the Supervisors meet Nov 1, but the potent influence of his his truly Christian spirit of, "Peace on earth, good will to men" will remain with them forever.

"Since 1871 he has made La Porte his home, happy and contented to live among those pine-clad hills and among those, who like himself, have found joy and contentment there. In the little mountain cemetary of that almost deserted village, where he chose to live, there they interred his frail and aged form. God in his goodness did not permit him to suffer on a bed of sickness. They found him reposing like one who wraps the drapery of his couch about him and lies down to pleasant dreams."

A host of old friends, among whom were more than 100 from Quincy, Johnsville, Blairsden, Howland Flats, Port Wine, Strawberrry Valley and other adjoining towns, gathered at his bier to pay their last sad rites. Of relatives there seemed to be none. The Masonic Lodge of which he was an old and respected member, had charge of the funeral ceremonies, with the Odd Fellows as an escort and the Native Sons as pall bearer.

"A.W. Keddie, who officiated, read the solemn burial service in a most impressive manner. A response was given by H. G.l Dorsch. M.C. Kerr acted as marshal.

"In behalf of the different organizations participating, J. E. Wilson extended grateful thanks to all who in any way expressed their kindly sympathy and assistance in the last ministrations of their departed friend. The occasion was unusually sad in that there was no relative to return to his home.

"All lies buried there."

Quicksilver Charley Hendel

References:
Interviews by the author
The Downieville Mountain Messenger
The Virginia City Territorial Enterprise
The Alpine Chronicle
The American Geophysical Union
The Sacramento Bee
The New York Sunday News, Nov. 6, 1949
The Virginia City Territorial Enterprise, Nov. 27, 1867
The Plumas Independent
The Nevada State Journal

a final look...

Longboards have a way of re-appearing through the decades.

1952

1941

Top photo, competitors at La Porte. Middle photo, long boards come back at Sugar Bowl in 1941. Bottom photo, Lost Sierra Snow-shoers gathered for a grand finale of the Snow-shoe Era in 1911.

1911

Appendix 1

"The palmy days of snow-shoe racing in Plumas and Sierra Counties...when the world was young and a delightful place, free from restraint and with enthusiasm to support new ideas." Quote attributed to Wendell T. Robie.

"a mountain to move!"

Skiing flourishes today in the Sierra Nevada thanks to the dedication of men who exhibited leadership in the 1920s. Men such as Wendell T. Robie, the first president of the Auburn Ski Club and Jerry Carpenter, a California State Chamber of Commerce executive who recognized the potential found in the development of skiing for the state of California. These people, backed by a dedicated membership of skiers, were responsible for the rebirth of organized skiing that had started over a century earlier in the Lost Sierra.

On February 27, 1929 a small group met in Auburn to discuss forming the ski club. The founding of the Auburn Ski Club led to the formation of the California Ski Association. Wendell Robie was the first elected president. The other first officers were Lew Volz, vice president; Harry Ricksecker, Secretary; Lane Calder, treasurer; Art Sather, Otis Owen and Dave Gordon, directors.

The growth of skiing largely happened because pioneering spirit, like that found in Wendell T. Robie, led a band of small clubs in Nevada and California to vie for the chance to hold the 1932 Winter Olympic tryouts at Lake Tahoe and the 1960 Winter Olympics at Squaw Valley.

The beginnings were humble, but of one mind. During the summer of 1929 organized winter sports had already begun in resorts throughout the length and breadth of the Sierra Nevada. Out of this came the growth of new facilities such as a 65-meter ski jump at Lake Tahoe which emerged under the guidance of the Lake Tahoe Ski Club's Norwegian veteran skiers.

The story of how skiing grew in the Sierra Nevada has an odd twist, however. Ironically, the organization of skisport in California was to a large extent the result of an effort to attract the Winter Olympics to California that failed.

Despite the efforts of the Winter Sports Committee of the California State Chamber of Commerce, the state unsuccessfully bid for the 1932 Olympiad Winter Games. Then, in 1929, the International Olympic Committee awarded the 1932 winter games to Lake Placid, New York. The California delegation had failed to convince the committee back east that California had anything more than palm trees.

Those interested in promoting skiing in the Sierra Nevada were determined to turn failure into a long term success. In 1930 the National Ski Association gave the sanction to the California State Chamber of Commerce to organize the California Ski Association. (Some 20 years later in 1949, the name was changed to the Far West Ski Association.)

Wendell Robie organized strong club support. He put together an impressive list of CSA charter member clubs of California and Nevada that included Tahoe Ski Club, Yosemite Winter Club, Truckee Ski Club, Nevada City Ski Club, Auburn Ski Club, Viking Ski club and the Reno Ski Club. The clubs bid for major national ski races to bring U.S. recognition to the superior winter snows and alpine skiing in the Sierra Nevada.

The newly formed California Ski Association was accepted by the National Ski Association as an affiliated division at the annual national convention in Chicago, December 14, 1930. At this convention California was awarded the 1932 National Ski tournament.

The first sanctioned divisional meet was in 1931. The opening cross-country race was won by Ludvig (Vicki) Hasher of the Viking Ski Club, Los Angeles. The jump was won by Sig Vettestad of the Auburn Ski club. The meet was billed as an Olympic tryout for the 1932 Winter Olympic games at Lake Placid, New York.

In 1959 Squaw Valley, eager to prove the calibre of its Winter Olympic runs, welcomed an international field of star skiers, as well as unknowns, to the North American Championships, thus setting the stage for the Winter Olympics. In 1945, the Reno Ski club secured Far West Ski Association sanctions for the Slide Mountain Silver Dollar Derby National Slalom Championships and the Snow-shoe Thompson memorial Cross-country Championships.

On the other side of the border, Reno, Nevada, a willing partner to the growth of skiing in the Sierra Nevadas, played a key role in bringing the 1960 Winter Olympics to Squaw Valley.

In 1955 a small item in the Nevada State Journal reported the International Olympic Committee was seeking sites for the 1960 Winter Olympics. George Wingfield Jr, a wealthy Reno sportsman, had read the article just before he ran into Bill Berry, whom he saw coming out of the Riverside Hotel's revolving door.

"Why doesn't Reno do something?" he asked Berry.

Anyone who knows Bill Berry can visualize the intense blizzard of publicity that followed, spurring Reno leaders to the limit.

Soon a Reno promotional team headed east with a bid. Reno's bid as it turned out was unsuccessful. However, the publicity triggered by Reno's efforts led to Squaw Valley's bid which was prepared by Alex Cushing.

To qualify with the Olympic Committee, the bid required endorsement by three local ski clubs. Hands joined across the Nevada-California state line. Endorsements came from Reno Ski club, Lake Tahoe Ski club and Squaw Valley Ski club.

The success of the 1960 Winter Olympics is now history.

Skiing introduced in the Sierra Nevada by miners in 1853, was finally ready to grow into a world class, multimillion dollar enterprise; as it had long been in Europe.

The growth of the Auburn Ski Club

The history of skiing in California was tightly intertwined with that of the Auburn Ski Club and grew in stature from the remote mining camps clustered around La Porte to the major Sierra Nevada ski complex that exists today.

The club started with the Canyon Creek location. This incubator of ski jumpers in California, the first hill of any considerable size to be completed in California, was designed by Otto Lirsch. The club's first ski jump, good for 50 to 60-foot jumps, was built at Carpenter Flat four miles beyond Canyon Creek and Baxter.

In 1930, the first major tournament was held in the State by the short-lived American Ski Association. Hans Haldorsen and Sig Vettestad, representing the Auburn Ski Club, were the only competitors in the amateur class to hold their jumps on the Tahoe hill. Membership in the Auburn Ski Club soared to over a thousand as skiers joined at 50 cents each.

Greater expansion and accessiblity had become a key concern to members. Toward that end, the club had searched the Sierra Nevada mountains, between Auburn and Summit for ski hill locations. Scores of miles were traveled on skis in winter and carefully examined by automobile and on foot in summer to locate a "Super Ski Hill" and to definitely locate the club on the very finest ski area in California easily accessible by the highway and railroad.

The move from Canyon Creek was made in 1933 to the Cisco ski grounds, across the Yuba River from Red Mountain, below the SP (Southern Pacific) tracks at Tunnel Mountain about two miles from Cisco Grove. The club built warming huts and a master ski jumping hill good for leaps up to 200-250 feet. Alongside the jump hill a haulback was built. For many years the Auburn Ski Club used the site to stage state, national and international ski competition.

The property covered 100 acres and extended for a half mile along the south bank of the Yuba River near the Auburn-Lake Tahoe Highway and Southern Pacific Railroad. It adjoined a government section of 640 acres to the north so the future was safeguarded with an unrestricted use for skiing on the 740 acres.

Cisco was abandoned when Interstate 80 cut through the club's ski jump outrun and parking lot. The new interstate proved a boon because it opened up the possibility of a new site discovered by Wendell Robie and Roy Mikkelsen as they hiked along the proposed I-80 right-of-way. In its history the Auburn Ski Club had moved, one step at a time, to higher elevations, towards the highest final location at Boreal Ridge.

An option to purchase the land from James Sheritt, a friendly sheep range owner, was negotiated. After World War Two, Auburn Ski Club directors co-signed a note with the bank to finance purchase of the property.

November 19, 1963, led by the club's jumping star, Orin Ellingsen, plans for a 231-acre snow sports public recreation area at Boreal Ridge were revealed. It would be the greatest new ski attraction in western America. The club's new ground stands directly on the snow crest of the Sierra Nevada. The site promised snow "first, last and best."

Boreal Ridge was in full swing when it welcomed skiers for the 1965-66 ski season.

In the intervening years the Auburn Ski Club's enthusiastic officers and young men had spearheaded every ski expansion drive in the Sierra. For instance, in order to secure funding for road opening programs the club had brought California Legislators to the snow-line.

Here we quote from Wendell T. Robie's "Half Century of California Skiing," written in 1978 to commemorate 50 years of the Auburn Ski Club.

...

"These Auburn Ski Club skiers were real salesman. Here was a mountain to move. Here would be the most vital moment in all of California skiing development. It was determined to secure votes by giving the Legislature a sample of open road travel which the ski club would provide. Senator Bill Cassidy of Auburn extended an invitation to the whole legislature to be guests of Auburn Ski Club at its ground in the mountains, Sunday, Jan. 18, 1931. ...The invitation was widely accepted.

"A motor caravan of 56 autos left Auburn in the dark at 6 a.m., that Sunday morning. After picking up the legislators at Sacramento they arrived at Cisco, the Auburn Ski club's ski grounds."

Another invitation, publicized at the same time for that same day, invited the public to a great "Free Ski Jumping Event" at the Auburn Ski Club grounds.

Legislators arrived first followed by an endless streams of automobiles. A traffic counter recorded more than 2,400 automobile which drove into that dead-end road and hemmed in the cars bearing the legislators. After the ski jumping and entertainment, this became the most monumental traffic jam in the history of the California Highway Patrol. The magnitude of the entanglement was convincing evidence, proving a passionate assertion held by the Auburn Ski Club and Senator Bill Cassidy that gas taxes paid on fuel as a result of this extra public travel would foot the bill for plowing the state highways in winter.

Thus, the Auburn Ski Club convinced legislators that skiing would produce enough extra gas tax revenues to pay for snow removal to keep highways open.

Sen. Cassidy's bill was approved. For its efforts, the Auburn Ski Club received a "Vote of Thanks", handprinted on parchment, bound in red silk and black leather and titled in gold leaf. Today this document is a prized exhibit in the William B. Berry, Western America Skisport Museum on I-80 at Boreal Ridge.

Appendix 2

"...Working with Bill Berry we want to help provide in California, an outstanding historical attraction which will demonstrate the contribution to resource development, help and offer wholesome, vigorous, good qualities in American life by the skiers of the Far West." Wendell T. Robie, President, Auburn Ski club.

Western Skisport preserved in history - the museum

Wendell Robie hoped to see the history of skiing in the Sierra recorded for posterity. On October 15, 1965, he advised the Far West Ski Association regarding a proposed museum.

His sentiments led to the club's historical program and construction of the William B. Berry Western America SkiSport Museum following the move to Boreal Ridge. The 231-acre acquisition of the Boreal Ridge land and completion of the Western America SkiSport Museum led to private, public, national and international recognition of the contributions made to skisport by the museum.

The museum exemplified the aims and objectives of the United States Olympic Association, Far West Ski Association, U.S. Ski Association and the Federation of International Skiing (FIS). This initial planning called for facing the entrance of the proposed museum toward I-80 with flagstaffs, flying the Stars and Stripes, the California Bear Flag, the Nevada Battle Born flag and the Olympic Flag. Later to be added was the 1976 Bicentennial Flag and the statue of Snow-shoe Thompson. There was emphasis that the Western America SkiSport Museum would always complement the National ski Hall of Fame and Museum, Ishpeming, Michigan.

Construction of the museum forged ahead in 1967-68, sparked by Roy Mikkelsen, two-time Olympic champion.

The museum was dedicated on Pearl Harbor Day, Dec. 7, 1969. It has grown in prestige and stature ever since. Discoveries of fine exhibits and the donation of Snow-shoe Thompson's mailbag and skis along with the support of the Sons of Norway, are just a few factors that have contributed to the museum's credentials.

The Auburn Ski Club owned the land at its Cisco Grove ski area, with its north-facing slope tucked under the Southern Pacific Railroad right-of-way.

After World War Two, Wayne Poulsen installed a rope tow at Cisco Grove, on the Auburn Ski Club's land. He called the enterprise "The Snow-shoe Corporation."

The Auburn Ski Club maintained a ski hut at Cisco. Around 1954-55 the hut was considered for a museum. That idea soon collapsed when the California Division of Highways projected the new I-80 Freeway right of way. It cut right through the base of the ski area, taking the ski jump outrun and the parking area.

The talk of a ski museum site then switched its focus to the Slide Mountain Ski Area, which had been named alternate site for the 1960 Winter Olympic Games. The State of Nevada had voted $200,000 to support the Olympics, which was used to build the Nevada Center at Squaw Valley and the "Olympic Chalet" at Slide. The 2,400 square foot building included a 16-foot-wide fireplace donated by Raymond I. Smith and photo murals of the snow country donated by prominent Reno sportsmen. The chalet was dedicated Dec. 20, 1959 by Nevada Governor Grant Sawyer. The $40,000 building served as a youth center for young skiers immediately following the Olympics. Later, it was proposed as a suitable location for a ski museum. Chalet contributors, which included Smith, Bew Fong, William Harrah, Wayne Poulsen, Jerry

Wetzel, Hal Codding, Gene Christensen, Mrs. Gladys Mapes and Bill Berry, liked the idea.

Then Clara Beatty, director of the Nevada Historical Society, wrote a November 18, 1965 letter to Wendell Robie, president of the Auburn Ski Club, endorsing the concept of a ski museum.

Then one day in the spring of 1965, a sign appeared on the Auburn Ski Club's Donner Pass ski area property at Boreal Ridge on I-80.

The sign proclaimed:

site for
AUBURN SKI CLUB HEADQUARTERS
and William B. Berry
HISTORICAL MUSEUM
of Western America SkiSport.

No one was more surprised than Bill Berry, as no one had said a word to him about it.

Bill immediately called his friend Roy Mikkelsen, ex-Olympian, ex-National Ski Jumping Champion, former Mayor of Auburn and a sinew of the Auburn Ski Club.

"Please tell me what this is all about and what I should do?" he asked.

"I'll try and find out. I'll take care of it," Mikkelsen said.

Wendell Robie, president of the Auburn Ski Club, had decided a ski museum should be built around the Berry's collection of Snow-shoe Thompson artifacts, snow-shoes and other treasures from the Lost Sierra.

The Boreal Ridge ski area construction started that summer.

By the fall of 1967, the foundations for the museum were poured with Roy Mikkelsen supervising the job. It was to be his last public service, as a few days later he suffered a fatal stroke.

Bill Berry then talked to Harry Rosenberry, Auburn Ski Club treasurer. He was assured all was going well and the club would complete the museum.

Bill Berry had a hand in the success of the Boreal site having reported the efforts of former ski champions Mikkelsen and Orin Ellingsen.

In 1963 after the ski club had taken possession of the Boreal Ski Area land, Ellingsen took an Auburn Ski Club promotional exhibit to the first annual California Winter Fair at the Claremont Hotel, Berkeley. Since the club had not reserved any display space, Ellingsen set up the Auburn Ski club exhibit in the hotel's stairwell.

The exhibit stole the show and was headlined in all the news reports of the California Winter Fair that year.

snow country

The frozen beauty of the Lost Sierra, the center of California's heaviest snows, meant mining towns were blanketed with up to 30 feet of snow, often up to the roof tops. Residents would have to burrow their way out.

View from Gold Lake road. Snow on the Sierra Buttes, towering above Sierra City, southern gateway to the Lost Sierra.

Picture taken in the late 1950's. Will Spencer, of Gibsonville, used beams to hold up his house against the heavy pressure of snow outside.

Right, snow tunnels at La Porte circa 1907-1911.

snow-shoe mail carriers tracking La Porte

Above, shown on her longboards in 1951, Elizabeth Merian had a sixth sense about deep snows.

When the snow was not too deep horses brought passengers and mail to La Porte. Above, stage passes near La Porte in the early 1900s.

FROZEN ON THE MAIL ROUTE

Death Overtakes a Carrier on a Snow-Covered Trail in the Mountains.

HIS COMPANION NEARLY SUCCUMBS.

Privations of a Party of Eleven Miners Who Attempted to Travel a Distance of Five Miles Through the Snow—Cutting the Road With Their Hands—It Took Them Sixteen Hours to Make the Trip.

NEVADA CITY, January 6.—Malcolm P. McLeod, a native of California, aged 28, and employed as a mail-carrier, was frozen to death this morning at two o'clock. Yesterday afternoon he and John Grissell, about the same age, started on snow-shoes to take the mail and express to Washington, eight miles distant, expecting to get there by dark. Each had a twenty-pound load.

Two miles below Washington McLeod began to fail. His companion spurred him on a mile further to Old Man Thompson's cabin on the river. Thompson said he had neither food nor firewood, and could not keep them.

They abandoned the snow-shoes, and Grissell carried and dragged McLeod half a mile and to within half a mile of town, but could not get him beyond there, as he, too, was rapidly succumbing. He went into town for relief, and did not get there till 1:30 this morning.

A party of citizens hastened to the relief of McLeod. He was still breathing when they reached him, but died before they got him to town.

The carriers had become bewildered and traveled in a circle many hours, though experienced snowshoers and familiar with the line of the trail.

Tales come from many points to the north

Above, mountain stage driving, long considered a male occupation, was second nature for Elizabeth Merian of La Porte, Plumas County. She travelled every day over some of the most historic, scenic and dangerous roads in the Sierra. Here Elizabeth Merian shows how deep the snow drifted during the winter of 1951-1952 on the road between Onion Valley and Gibsonville enroute from Quincy to La Porte.

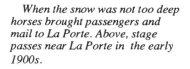

Above, snow scene at American House. Mail carrier Elizabeth Merian, left, in 1950 as she reaches out for mail delivered by carrier Dave Hall en route to La Porte. Both played key roles in the revival of the Alturas Snow Shoe Club started in 1867.

Fashions through time

Top. Skiers in 1914 gathered at Revelstoke, British Columbia. Photo shows women skiers wearing ankle-length skirts and long sweaters.

Bottom. Dressed for skiing in the Snow-shoe Era, La Porte, circa 1910. The ladies and their families readily adapted to the snow-shoe life. Photo above includes prominent townspeople, like the Cayot family, owners of the Union House hotel. From left, Mrs. York, Elizabeth Maxwell, Virginia Buckeley, Bill Merrill, Rose McIntosh Merrill, Elsie Squire Smith, Emma McIntosh, Hattie Mullen Fitzgerald, Annie Williams, Claire Cayot O'Rourke, May Schubert, Mel McIntosh, Mrs. Mel McIntosh, Annie Greeley and an unidentified woman on the far right. Seated from left, the partially hidden woman with the plaid skirt is Eva Cayot, then Elizabeth Robinson Merian, later to become La Porte's mail carrier; Pearle Robinson, Nell Corbet Kingdon and Isabelle Bustillos. Photo from Elizabeth Merian collection.

Fashions through time

Women wore jodphurs, pants, ski socks and legging for skiing as shown in this 1920 photo taken at Steamboat Springs, Colorado.

Auburn Ski Club's Sigrid Lamming, left, 1932 National Ski Tournament cross country champion at Lake Tahoe, displayed a snappy outfit.

Women flocked to the Truckee/Lake Tahoe ski slopes in 1932, drawn by National Ski Tournament publicity. Woolen ski sweaters and pants became popular, mixed with varied accessories.

*the races:
when dope was king*

Repeat winner, Sugar Bowl, 1941, Johnny Redstreake in lead, successfully defended his 1938 Johnsville title. Johnny Redstreake used Ab Gould's "Lightning Dope" to win the Snow-shoe Renaissance Race. He split the purse with Gould.

Johnny Redstreake

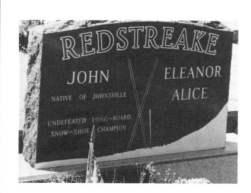

Lower left, Johnny Redstreake's grave at the Johnsville Cemetary. Above, the 1938 Johnsville snow-shoe race champion holds his "lightning doped" shoes. He is congratulated by Monte Sobrero, left, third place winner.

the races: when dope was king

Albert (Ab) Gould, 63, watched by his brother, Art, carefully applied "lightning dope" to his snow-shoes before defeating Reno Champion, Wayne Poulsen, at Johnsville in 1938. His "dope" on Johnny Redstreake's skis made a winning combination.

Alturas Snow-shoe Club sponsored 1938 Snow-shoe Race Renaissance.

Left, Lenny O'Rourke bangs on tin can signalling race start. Middle, Ab Gould, winner, gets set to beat the latest generation of skiers. Bottom, a demonstration of Swiss flying kilometer skis, proving them no match for snow-shoes.

Andy Modglin, oldtime "dope" maker and Sierra County Supervisor, lets 1951 snow-shoe race winner, John Cowley, sniff his 93-year-old dope recipe.

the lost sierra revisited

From left, Bill Berry, Pat O'Kean, Harry Rosenberry, Wendell Robie gather around the grave of Gold Rush miner William Wier which was restored at Poker Flat in 1971.

Wearing the familiar Alturas Snow-shoe Club star, author Bill Berry, left, toasts 1951 Snow-shoe race winner at Reilly's Bar, La Porte.

Welcomed under a backdrop of snow-shoes in 1949, Ellin Mackay Berlin, center, is greeted by Marguerite Delahunty, "The Balsam girl" of the Snow-shoe Era, left. Author Bill Berry, right.

The author stands in the doorway of what was left of A.T. Merian's American House, July 28, 1989, now suffering neglect beside the paved highway between Marysville and La Porte.

the lost sierra revisited

Author with Mrs. Truman (Helen Weaver) Gould, on the porch of her La Porte home, July 28, 1989. CW photo.

It was here at the Four Hill mines, Johnsville, that author Bill Berry first met and talked with a snow-shoer in 1931. This meeting resulted in a growing fascination with Lost Sierra snow-shoeing.

Celebrating the Sierra Nevada Ski Centennial 1854-1954, Reno, March 27-28, 1954, were Snow-shoe Era descendents of the Lost Sierra pioneers, introduced at the Riverside Hotel by the author at the microphone. From left, Johnny Redstreake, behind Bill Berry; Birdie Haun Swingle, Andy Swingle, Will Primeau, A.T. (Jumbo) Merian, Mrs. Elizabeth Merian and Mrs. Ollie Pixley.

william b. berry western america skisport museum

The Bicentennial flag is raised in April 5, 1975 at the Western America SkiSport Museum.

Boreal Ridge and the museum in 1969, already becoming a worldwide ski attraction.

---

exhibiting the life of snow-shoe thompson

Irene Simpson Neasham, president of the California Historical Society, donated Snow-shoe Thompson's mailbag and snow-shoes, from the Wells Fargo Historical room.

Left to right, Harold Moore, Stanley Moe and A.J. (Mac) McCabe, representing Lodge No. 78, Sons of Norway, Yuba city, attended the May 15, 1976 recognition of Thompson statue by Norwegian Ambassador, Soren Christian Sommerfelt, right.

museum records organized skiing

Albert Sigal, San Francisco, 1954 President (USSA) U.S. Ski Association, brought recognition to California skiing.

Jerry Carpenter, director Winter Sports department, California Chamber of Commerce. Encouraged Auburn Ski Club.

Wendell T. Robie, Auburn ski club President, first president and founding member of California Ski Association, now (FWSA) Far West.

Roy Mikkelsen, two time USSA Ski Jumping champion 1933 and 1935, put California skiing on the map.

Western America SkiSport Museum, Feb. 19, 1972, was marked a Nevada County point of historical interest at a presentation by Wendell T. Robie, left foreground, Auburn Ski club president and Mrs. Leland Lewis, center foreground, Chair, Nevada County Landmarks Commission.

Harry Rosenberry, Auburn Ski Club Treasurer, left, talks to John Watson (FWSA) Far West Ski Association Historian at Museum, May 15, 1976. Dedication of Snow-shoe Thompson statue.

Norway's King Olav V, right, October 23, 1975. At San Francisco, accepted honorary membership in the U.S. Ski Association, From left, then USSA president Richard Goetzman, Alec Davidsen, and the author.

Index

A
Abbott, Bonnie, 67
Abbott, N.P., 100
Accord, Art, 199
Ah Pock, 219
Alleghany, 143
Alleghany District, 141-144
All Fool's Day, 41
Alley, Dick, 69
Alpine County Chronicle, 84, 86-94
Alpine Meadows, 206
Alturas County, 41
Alturas Odd Fellow's Lodge, 166
Alturas Ski Club, 41
Alturas Snow-shoe Club, 40, 41, 58-64, 67-69, 84-88, 96, 98, 100, 107, 120, 159, 162, 166, 190, 205, 212, 216, 219-222, xi
American House, 17, 219, vii
American Ranch Hotel, 128
American Ski Annual, 220
American Ski Association, 202
American Valley, 126, 127
Auburn Ski Club, 78, 83, 185, 200, 201
Aurora Ski Club, 215, 217

B
Babb, Mrs. George, 157, 158
Bain Escape Map, 22
Bain, Miner, 22, 23
Baker, Col. E.D., 143
Bald Mountain, 143
Bald Mountain Mine, 70
Baldwin, Beach, 196
Baldwin, E.H., 196
Barnes, Charles W., 119
Barnes, B.W., 60
Barnes, Benjamin Wilcox, 106, 166, 168, 174
Barnes, Charles, 65
Barrett, J., 41
Barrett, John G., 101
Barrett, John H., 100
Beach, Rex, 200
Bean, Abraham Lincoln, 162
Bean, Joel, 162
Bean, John, 100, 162
Bean, Mrs. 163, 164
Bean, Vernon J., 162
Beckwourth, Jim, 20, 27, 184
Beckwourth Pass, 20, 29, 184, 186, 223
Beebe, Dan, 213
Bengter, Stig, 84
Beresford, Husky, 198
Berlin, Irving family, 223, 224
Berlin, Ellin Mackay, 223
Berlin, Mary Ellin, 223
Berman, John, 41
Berry, Davy, 219
Berry, Samuel, 114, 115
Berry, State Senator Swift, 75, 79
Berry, Wallace, 199
Berry, William B., 38, 79, 83, 216, 217, 218, 220, xii, xiii, xiv, see also William B. Berry
Bicentennial Celebration, 95
Bidwell's Bar, 17, 20, 22, 33
Bishop, Fred, 75
Black, Jacob, 111
Blatt, Bob, 202
Blatt, John, 202
Bliss, D.L., 196, 197
Bliss family, 197
Bliss State Park, 196
Blodger, Andy, 83, 93

Blodger, Olaf, 83
Bona, Signor, 157, 158
"Bonanza John," 143
Booth, Governor Newton, 121, 124
Boreal Ridge, 38, xiv
Borell, Bedrock, 154
Bosworth, Hobart, 199
Bowman, Bill, 151
Bowman, Dee, 151
Bracken, Carl, 112
Brewster, Doc, 87
Brewster, S.T., 119
Brown, George, 66, 70, 100
Brown, Lily, 65
Brownsville, 143
Bryant, Dr. Edmond, 223, 224
Buckbee, John R., 121
Buckeye, 33
Buck's Ranch, 17, 22, 23
Buek, Dick, 202
Burelle, George, 97
Burelle, Jerry, 223, 227
Burelle, Medrick, 223
Burke, John, 149
Bustillos, Coach Jesus
Bustillos, J.J., 39, 50, 97, 101
Bustillos, Joe, 101, 103, 137
Bustillos, Medev, 61, 107
Bustillos, Mike, 39, 101, 137

C
California Illustrated Magazine, 27
California Mining and Scientific Journal, 37
California Ski Association, 198, 210, 203, 205
California State Chamber of Commerce, 200, 222
California State Library, 43
California Volunteers, Company F, Fifth Infantry, 24
Camptonville, 141
Capitol Hotel, 128, 144
Carpenter, Jerry, xv
Carson Valley, 74
Cathay Circle Theater, 81
Carvin, Elton O., 144
Cassidy, Bert A., 199, 201, 202
Cassius, Augustus, 135
Cavelier, Robert, 27
Cayot, Ray, 97, 219
Central Pacific Railroad, 60, 90, 176-182, 185,192, 195, 196,
Chadwick, 55
Challenge, 142
Chambers, S.C., 187
Chateau Frontenac, 198
Childs, Judge John S., 81
Chilean miners, 130, 132
Chinese laborers, 176, 177, 195
Chorpenning, George, 74
Christopher, Louis, 134
Church, Abram D., 114
Church, Dr. James Edward, 203-206
Cisco, 177, 180, 181, 187, 188-191
Cisco Grove, 90, 185, 201
Civil War, 24
Clark, Charles, 65
Cleveland Gold Mining Co., 162
Clinch, Bill, 117
Clinch, John, 119
Clinch, William, 48, 49
Clough, G.G., 121
Coats, J.P., 135

Codding, Hal, 202, 213
Colfax, 177
Collins, James B., 138, 139
Colt, Ben, 29
Comstock Lode, Nevada, 76
Condon, Mary, 113, 114
Congressional Record, 79
Conlon, D., 119
Conly, John, 34, 60, 62, 119, 120, 127
Cook, George E., 133
Corbett, David, 105
Corbett Hotel, 161
Corbett, John, 170
Cosker, 219
Cosker, Bud, 213
Cosker, James, 100
Cosker, Michael, 151
Costello, Tom, 100
Cowley, John, 223, xi
Cozzalio, Charles, 203
Crabtree, Lotta, 156-161
Crabtree, John 156-158
Crabtree, Mary Ann, 156-158
Crandell, Joe B., 80
Crawford, James, 175
Cray, L.D.C., 60
Creed, Haymond, 60
Crew, Alex H., 34, 60, 119, 127, 170
Crocker, Col. Charles W., 53
Crocker, Charles, 176, 177, 178
Crystal Peak, 177, 187
Curtis, Jerry, 127-129

D
Danfer, Edward, 41
Danforth, Ed, 52, 105
Dalton, Father Thomas, 168
Darley, Rev. George M., 167
Daugherty, Henry, 163
David, Bill, 198
Davidsen, Alec, xv
Davis, George A., 175
Delahunty, Father, 149
Delahunty, Frank, 164
Delahunty, Henry, 163
Delahunty, Marguerite M., 164, 224, xii
Delahunty, Millie, 117
Delaney, Frank, 151
de la Salle, Sieur, 27
Den Norske Turn og Skiforening, 217
de Quille, Dan, 75, 77, 80
Des Baillets, Ernest, 197, 205
Delta Saloon, 155
Devlin, Jerryann, 220
Dewey, Albert T., 173
Diamond Valley, 76, 79
Diamond Valley Ranch, 80
dog sleds, 33
Donner Lake, 176, 177, 179, 184, 195
Donner Lake House, 182
Donner Party, 7
Donner Pass, 29, 176, 180-185, 210
Donner Summit, 176, 186
dopemakers, 39, 40, 48, 57, 96,97, 103, 104, 225
Dorsch, H.G., 226
Dorset, Henry, 198
Douglas County Ski Club, 82, 83, 93, 197
Downey, Major William, 34
downhill skiing, 30
Downieville, 15, 25, 34, 47, 140, 141, 209
Downieville Clampers, 171

Downieville Mountain Messenger, 25, 31, 45, 41, 50, 52, 116, 156, 162, 171, 172, 186, 196, 225
Drake, H.G.O., 170
Dressler, Albert, 11, 26, 27
Driver, Tom, 25
Dugan, Jim 150
Dunlap, Mordicain, 22
Dunlap, Rev. R.R., 168
Dubuque, 219
Dubuque, Al, 101
Dutch Flat, 177, 195
Dutch Flat swindle, 190
Dutchman's Ranch, 100

E
Eagle Ski Club, 211, 212
Eastland, Bill, 111
Ebbetts Pass, 5
Echo Summit, 203
E. Clampus Vitus, 69, 76, 80, 105, 166, 170, 171
Edmunds, Earl, 211, 212
El Dorado Saloon, 18
Ellingson, Ron, 223
Emigrant Gap, 196
Ermantinger, Henry, 100
Espinal, Martin, 83, 93
Eureka Peak, 2, 6, 9, 10, 130
Evans, Col. Albert S., 53
Evenson, Larry, 198
express snow-shoe carriers, 32, 33

F
Fair, James, 147
Far West Ski Association, 220, 222
Feather River, 2, 5, 6, 19, 20, 21, 126
Feather River Highway, 225
Feather River Inn, 130
Feather River Express, 33
Federation Internationale de Ski, 174
Felton, Charles N., 143
Fennimore, Tom, 161
Fiddler's Flat, 23
Field, Stephen J., 143
Fink, Henry, 52
Flood, James, 147
Flournoy, John, 100
Forest City, 141, 143
Forty-niners, 3
Four Hills mines, xiii
Francis, Robert, 63, 64, 67, 116-124, 219
Francis, William, 67, 99
Franz, Lin, 100, 103
Freeman, Dr., 112

G
Gaiennie, Frank, 203
Galbe, Jorgen, 83, 85, 94
Gallagher, Reddy, 100
Galloway Hill, 25, 141
Gangloff, George, 52
Garvis, Jann, 62
Gatiker, John, 113
Geart, District Attorney H.L., 121
Genoa, 74, 76, 81, 95, 219
Gentry, Rev. J. C., 169
ghost towns, 3
Gianotti, John, 84
Gibb, Thomas, 105
Gibson Dramatic Club, 70
Gibsonville, 4, 15, 29, 34, 36, 49, 142, 151-154
Gibsonville Boys, 155
Gibsonville Herald, 173
Gibsonville Snow-shoe Club, 122, 221
Gilbert, Dale, 83, 93

Gish, Lillian, 199
Glenbrook, 196
Goetzman, Richard, xv
Gold Lake, 1, 15, 18, 19 (map), 20
Gold Mountain, 130
Golden Spike, 189, 191
Goodwin, J.D., 121
Goodyear's Bar, 25, 141
Gordon, Rev. William, 34
Gore, D., 170
Gould, Albert (Ab), 101, xi
Gould, Art, 214, 219
Gould, Bill, 102
Gould, Helen Weaver, 11, 208, xiii
Gould, Jake, 142, 214
Gould, Truman, 44, 214
Granville, Zachariah, 219
Grass Flat, 161
Grass Valley, 158
Gregg, Jacob, 29
Gregory, Jackson, 200
Grinden, Harold A. 212, 115-218
Grissel, John, 112, 113
Grizzly Mine, 159
grooving tool, 48
Grove, Rev. Philetus, 169

H
Hackett, W.H., 175
Haffey/Cosker home, 149, 154
Haffey, Elizabeth C., 154
Haffey, John J., 151, 152, 154
Haffey, Len, 149, 151, 155
Haight, Gov. H.H., 121, 189, 190
Haldorson, Hans, 200
Halley, Ellen, 150
Hall, Dave, 219, viii
Hall, Doc, 119
Hangtown, 5
Hansen, Charles, 134
Harold, John, 114
Harris, 101
Harris' store, 5
Hart, William S., 199
Harte, Bret, 116, 148
Haugen, Lars, 197
Haun, Anna Birdina, 128, see "Birpie Swingle"
Haymond, Creed, 60, 120, 121, 170, 219, 224
Haymond, W.C., 121, 122
Head, Clara, 187
Hearst, George, 147
Heath, Clarita, 202
Heavenly Valley, 203, 206
Helm, Alfred, 173
Hendel, Charles W., "Quicksilver Charlie," or "Wood Box Hendel," 36-44, 52, 53, 84, 85, 104, 126, 160, 168, 170, 171, 172, 220 225-226
Henness Pass, 29, 141, 147, 195
Hewitt, Andy, 102, 154
Hewitt, Charlie, 100
Hildebrand, Dr. Joel, 202
Hillman, 219
Hillman, Bill, 101
Hillman, George, 101
Hillman, Lester, 97, 101
Hillman, Louis, 101
Holcomb, Alice Pike Steward, 98, 99
Holcomb, Thadius, 98
Holmenkollen Cup, 84, 94
Hope Valley, 74
Hopkins Forks, 49
Horn, John, 29
Horn, William, 29, 52
horse snow-shoes, 33, 34
Howard, Dr. Frank, 53

Howe, A.J., 70, 100
Howell, William, 61
Howland Flat, 36, 41, 43, 45, 50, 97, 142, 148
Howland Flat Ski Club, 41
Howland Flat Snow-shoe Club, 40, 43
Hudson's Bay Company, 29, 209
Hungerford, Marie, 223
Hunsinger, Fannie Woodward, 105, 134, 137
Husker, Gabe, 55, 56
Husum Hotel, 46, 47
Hydraulic mining, 155

I
International Olympic Committee, 197, 198
International Ski Federation, 201, 205
International Snow and Ice Conference, 206
Irish, John, 29
Iseman, N.B., 119, 217
Ispeming Ski Club, 218

J
Jamison Ridge, 201, 211
Janss, Bill, 202
Jedkins, Silas, 99
Jersey Flat, 25
Johnsdotter, Gro, 75
Johnson, J.A., 143
Johnson, Martin, 200
Johnsville, 1, 15, 136, 139
Johnsville Boys, 100, 134
Johnsville Snow-shoe Club, 221
Jones, 220
Jones, Benny, 67
Jones, James, 168
Jones, Tom, 138
Jordon, Francis, 125
Joy, Lottie, 63, 65, 159
Judge, Matthew, 99
Jump, Dr. Alemby, 109
Junction Bar, 33

K
Keddie, A.W., 226
Kelleghan, Patrick, 100
Kellog, William W., 172
Kelly, 220
Kelly, Elza, 49, 117, 122
Kelly, Henry, 63
Kelly, Mike, 65
Kelly, Tom, 41
Kenn, Alex, 117, 122
Kenny, Jack, 169
Kenny, John, 154
Kerr, M.C., 226
Keystone Mill, 146
Keystone Mine, 110
Kidd, Billy, 223
Killey, J., 100
Kingdon, R.H., 100
Kit Carson Pass, 5
Kleckner, Abe, 119, 220
Knight, Mrs. Laura, 201
Kongsberg, 77
Kyne, Peter B., 200

L
Ladies of Lost Sierra, 63
Larison, Jim, 219
Lake House, 192
Lake Placid, 198
Lake Tahoe Ski Club, 201
Laming, Sigrid, 207, ix
Langley, Roger, 222
Langhorst, Lillian, 138
La Porte, 3, 5, 9, 11, 13, 30, 34, 35, 38, 44, 47, 58, 68, 142, 148, 158, 206, 219, vi, viii

La Porte Hotel, 66
La Porte Masonic Lodge, 170
La Porte Snow-shoe Club, 60, 127
La Porte Track, 90
La Porte Union, 92, 93, 96, 106, 172
Larimore, Tom, 39
Larison, Jim, 219
La Salle, John, 27, 28, 29
Lassen's Horn, 12
Last Chance Valley, 15
Lee, Frank, 46, 49, 50
Legan's Saloon, 59, 120
Lewis, Mary Ann, 161
Lewis, Mrs. Leland, xv
Lexington Hill, 15, 137, 219, 221, 222
Liberty, O., 100
Lie, Bjorn, 83, 93
Littick, Charley, 66, 70, 97
Little Grass Valley, 26
Lloyd, Kenny, 202
Lloyd, Mary, 65
London, Jack family, 199
longboards, 36, 38, 227
Lost Sierra, 2 (map), 6, 13, 53, 108, 131, 216,
Lost Sierra Rainbow, 7
Lowry, C., 170
Lucky Boy mine, 213
Lynch, Judge, 53
Lyons, Jimmy, 149

M
Mack, Effie Mona, 83
Mackay, Clarence, 224
MacKay, Ellen, 223-224, xii
MacKay, John W., 143-147, 196
MacKay, Marie Hungerford, 223
Madden, John, 100, 152, 155
Madden, M., 22
mail carriers, 3, 31, 32
Mangseth, Ron, 223
Marakeet, 132, 133
Marakeet's Bordello, 133
Marysville, 17, 161
Marysville Appeal-Democrat, 214, 215
Mason, 202
Mason, Clint, 185
Mason, Henry Harrison, 27
Mason, John T., 27, 34, 113
Mason, T., 100
Mason, William, 27
Masonic Lodge, 61, 166, 226
Matteson, T.J., 76
Maynard, Wilbur L., 198, 205
Maxom, Jess, 203
McCabe, Mac, 79
McClare, Father, 169
McClatchy, James, 191
McDonald, Charles, 134
McDougall, William, 62
McDowell, D., 62
McGee, Alexander, 112
McGlashen, Charles, 199
Mckay, Earl, 203
McLaughlin, 220
McLeod, Malcolm, 112
McLean, Mr., 135
Meadow Lake, 180, 181,
Meadow Lake Sun, 187
Meadow Valley, 17, 22, 33, 34
Meffley, Alice W., 105, 138
Mendenhall, Pop, 29
Merian, A.T., xii, xiii
Merian, Elizabeth, 46, 219, 221, vii, xiii
Merrill. W.R., 81
Messelt, Carl, 82, 83
Metcalf, Bill, 99, 116, 122, 123

Metcalf, Eliza Kelly, 119
Metcalf Hotel, 119
Metcalf, William, 46, 48-51, 62,
Methodist Episcopal Church, 34, 166
Michigan Daily Mining Journal, 218
Mighel, 220
Mikkelsen, Roy, 77, 78, 202, 223, xv
Miller, Joaquin, 116
Miller, Philip, 128
Miller, Robert, 209, 210
Mills, Darius Ogden, 147
mining companies, 114, 143
Mitchell, Dick, 202
Mix, Tom, 199
Modenese, Dorothy, 221
Modglin, Andy, xi
Moe, Stanley, xiv
Mokelumme Hill, 105
Montez, Lola, 157, 158
Moore, "Tin Plate", 201
Moore, Harold, xiv
Morgan, Amanda L., 162
Moriarty, John, 102, 154
Moriarty, Margaret Ellen, 138
Morrison, Sanford, 114
Morristown, 34
Mount Lola, 157, 158
Mount Pleasant, 40
Mount Rose, 204, 206
Mullen, A.L., 173
Mullen, Ed, 39, 101
Mullen, Col. James F., 100-103, 212, 215-218
Mullen, Ira, 46
Mullen, Joe, 137
Mullen, Neal (Buck), 39, 137, 154, 169, 220, 221
Murphy, William, 191
Murray, James, 11, 26, 219
Murray, Michael, 147

N
Nash, Vicki, 95
National Ski Association, 174, 198, 210, 212, 217, 220, 222
National Ski Jumping Championship, 198, 201
Native Daughter's of the Golden West, 128
Neasham, Irene Simpson, xiv
Ned, Little, 143
Neidlinger, Sally, 221
Nelson, Charles, 178
Nelson Creek, 23
Nelson Point, 9, 15, 18, 127, 139
Nevada City, 110, 115
Nevada City Enterprise, 184
Nevada City Transcript, 184
Nevada City Snow-shoe Band, 111
Nevada Comstock Lode, 76, 141
Nevada Ski Centennial, xiii
Nevada Winter Carnival, 205
Newark, 142
Nibecker, Mrs. John, 114
Norheim, Sondre, 69
Norman, Napoleon, 48, 63
North Star-at-Tahoe, 206
Norway, Telemarken district, 73
Norwegian Consul General, 83
Norwegian Embassy, Washington, D.C., 73
Norwegian model skis, 26
Norway skate, 26, 37
Norwegian snow-shoes, 3, 11, 32, 47, 51, 53

O
O'Brien, Jack, 146, 147
O'Brien, James, 29
O'Brien, William, 147
Odd Fellows, 69, 166, 168, 169, 170

O'Donnell, Father, 169
O'Hanrahan, Tim, 201
O'Kean, Pat 149, 150, 151, xii
Olav V, King of Norway, xv
Old Black Dope, 103
Old Club Foot, 115
Oliver, Robert, 48, 49, 61-64, 70, 99, 116-120
Olsen, Tosten, 75
Olympic Games, 198
Olympic Hill, 198
Onion Valley, 3, 9, 15, 20, 28, 30, 33, 219 38, 46, 47, 49
Onion Valley Snow-shoe Club, 221
O'Rourke, 219
O'Rourke, Cleveland, 35, 101
O'Rourke, Clyde, 61, 101, 107
O'Rourke, Leonard (Lenny), 103, 128, 129, 146, 211, 212, 219, 224, 225, xi
Oroville, 49
Oroville Mercury, 213
Osborne, D.H., 66, 92, 171
Osborne, Dennis, 221

P
Pacific Coast Intercollegiate Ski Union, 205
Page, Anita, 201
Passeta, Billy, 138
Passinetti, Pete, 203
Peavine, 17, 22, 23
Perkins, Dean, 220
Perkins, William, 113
Peterson, Harry C., 43
Peyton, Larry, 199
Pezzola, Steve, 130, 132
Pickford, Mary, 199
Pierce, George, 75
Pierce, Rev. G.C., 170
Pike, 220
Pike, Alice, 98-100
Pike, C.E., 107
Pike, Charles, 98
Pike, Ed, 101
Pike, Harold, 61, 103, 107
Pike, Joe, 101
Pike, Will, 43
Pilot Peak, 9, 28, 36, 91, 126
pilot plow, 189
Pine Grove, 36, 40
Pine Grove Snow-Shoe Club, 41
Pixley, Mrs. Ollie, xiii
Placerville, 219
Placerville Ski Club, 202
Plumas Argus, 32, 131, 174
Plumas Boys, 90, 84, 130
Plumas County, 23, 24, 43, 217
Plumas County Chamber of Commerce, 222
Plumas County Library, 44
Plumas County Museum, 128
Plumas County Snow-shoe Club, 221
Plumas-Eureka State Park, 10, 126, 130
The Plumas National, 174
Plumas National Forest, 19
The Plumas Standard, 174
Pocahontas Mining Company, 52
Poker Flat, 15, 97, 142, 148, 219
Poker Flat Snow-shoe Club, 221
Pollard, John G., 48, 62, 120
Poorman's Creek, 18, 36, 38, 46, 47
Poorman's Diggings, 30
Porter, John, 47
Port Wine, 142
Port Wine Snow-shoe Club, 64, 65, 69, 221
Post, Dodie, 202
Poulsen, Wayne, 204-206, 211-213, 222
Poverty Hill, 15
Primeau, 219

Primeau, Al, 61, 101, 107
Primeau, Bill, 50,
Primeau, Elmer, 43
Primeau, Will, 61, 107, 221, xiii
Promontory Point, 195
Putah, Creek, 74

Q
Quicksilver, 43, see also Hendel
Quincy, 15, 142

R
Rabbit Creek, 1, 11, 18, 26, 27, 29, 30, 34, 53, 172, 173
Rabbit Creek House, 29
Railroad Hotel, 90
Ralston, William, 148
Ramsey, Allen, 137
Ramsey, Bobby, 137
Ryan, Michael, 113
Reddish, Jack, 221
Redstreake, Johnny, 208, 213, 220, 222, x, xii
Reno Ski Bowl, 220
Reno Ski Club, 85, 201, 220
Rich Bar, 15, 21 (map)
Richmond Hill, 49
Riendeau, Peter, 48, 70, 99
Ritchie, Bob, 143
Riverside Hotel, 222
Robie, Wendell T., 38, 198, 201, xii, xv
Robie, Mrs. Wendell T., 200
Robinson, Bill, 69
Robinson, Sylvia, 164
Robinson, George R., 114
Rodolph, Katy, 202
Rolph, Gov. Sunny Jim, 201, 202
Rose, Col. R.H., 29
Roseberry, Harry, xii, xv
Roseville, 177
Ross, Alex, 99
Ross, William, 99
rotary plow, 192
Roubley, George, 61, 63, 69
Rowley, Richard, 83, 93
Royal Arch Masons, 170
Rubicon Point, 196
Rudolph, Katy, 202

S
Sacramento, 177, 186, 195
Sacramento Bee, 191
Sacramento Daily Union, 40, 89, 182
Sacramento Pioneer Society, 190
Sampell, Margaret, 138
Saulter, Frank, 63, 99,
Saulter, Fred 121, 122
Saw Pit Flat, 46-48, 99, 116, 118
Sawyer, Elliott, 205
Scales, 142
Scales camp, 17, 162
Schooly, Thomas, 22
Schoonmaker, Frank, 175
Schonnars, Dave, 25
Schram, John, 52
Schroll, Hannes, 220
Schubert, Elmer, 164
Schwarzenbach, Chris, 202
Scientific Press, 225
Scott, 220
Scott, Bob, 155
Scott, Charles A. Sr., 100, 119, 146, 150
Scott, Charles Jr., 155
Scott, Charles Reid, 150
Scott, Gwen Ramsey, 105
Scott, Walter, 155
Serch, Charles, 170

Sexton, Judge Warren Key, 211
Sharon, William, 147
Shield, Fred, 222
Shingle, Warren, 210
Sibbly, Henry, 67
Sibbly, Lon, 102
Sibbly, Louis, 100
Sierra Boys, 84, 86, 87, 90
Sierra Buttes, 9, 17, 110, vi
Sierra Buttes Inn, 144
Sierra City, 17, 140, 141, 145
Sierra County, 46, 47, 140
Sierra County Blues, 170
Sierra County Tribune, 174
Sierra Loop Ski Center, 206
Sierra Nevada Ski Centennial, 220
Sierra Ski Ranch, 206
Sigal, Albert, 220, xv
Silver Mountain, 72, 73, 77, 84, 86
The Silver Platter, 223
Singleton, Agnes, 73
Sisson, James, 77
Ski Cradle, 216
Ski Events, 83, 201-203
ski tournaments, 201-203
ski wax, 39, 40
Skinnerland, Einar, 83, 93
Slide Mountain, 206
Slyter, Robert I., 112
Smith, Jedediah, 209
Smith, Raymond, 85,
Smith, Verda, 164
Snowbird Club, 197
Snow Carnival, 201
snow manufacturing, 47, 98
Snowshoe Era, 3, 8, 30, 44, 69, 128, 145, 210, 216-223
Snow-shoe Flat, 1, 46, 48
Snow-shoe women, 47, 54, 63, 151, 156, 158-165, viii, ix
snow-shoe mail carriers, 33, 34, 112
snow-shoe races, 30, 36, 37, 38-42, 48, 51, 64, 66, 145, 172, 187
Snow-shoe Thompson,
see Thompson, John A.
Snow-shoe Thompson Monument, 83
Snow-shoe Thompson Race, 85
Sobrero, Jon, x
Sobrero, Monte, x
Soda Bar, 15, 22
Sommerfelt, Soren Christian, 79
Sommerseth, Leif, 84
Sonora Pass, 5
Sons of Norway, 79
Southern Pacific Hotel, 198, 199
Southern Pacific Railroad, 191, 192, 197
Spanish Diggings, 29
Spanish Flat, 18
Spanish Ranch, 17, 22
Spencer, Elizabeth, 149, 151-153
Spencer, William H., 148, 151-154, vi
Spooner's Summit, 93, 197
Sports Illustrated, 98, 216
Squaw Valley, 202-205
Squaw Valley-Lake Tahoe Ski Club, 202
Squires, Elias G., 62, 66, 99
Squires, Eugene, 99
Squires, Orville, 70, 101
Stanford, Leland, 176-178, 188
Starr, Hattie, 63
Sterling, Ford, 199
Stevenson, Tom, 99
Steward, Frank, 3, 8, 97, 98, 100-104, 225
Stewart, Frank, 73
Stewart, Senator William M., 79, 142, 143
Stoddard, J.R., 12, 13, 14, 15, 118

Stone, William, 99
Stout, H.P., 100
Strawberry Valley, 17, 20, 161, 162, 219
Street, Charles, 105
St. Clair, James W., 170
St. Louis, 36, 40, 142
St. Louis News, 173
St. Louis Hotel, 105
St. Louis Snow-Shoe Club, 41
Strobridge, James Harvey 187, 195
Sugar Bowl, 204, 206, 220, 222
Sutter's Fort, 12
Sturgis, William, 61
Summit Hotel, 192
Sweeney, Daniel, 148
Swett, R., 175
Swigart, Pop, 219
Swingle, Andy, 129, 220, 221, xiii
Swingle, Birdie, 126-129, 221, xiii

T
Table Rock, 9, 148
Table Rock Hotel, 163
Table Rock Snow-Shoe Club, 45, 66, 67, 90, 92, 155, 162, 171, 221
Tahoe City, 184
Tahoe-Donner, 206
Tahoe Tavern, 190, 197
Talbitzer, William, 152
Taylor, Mart, 157
Territorial Enterprise, 187
Thistle mine, 4
Thompson, Don, 223, 227
Thompson, John A. "Snow-shoe," 5, 69, 72-81, 83, 84-95, 98, 138, 142, 178, 181, 185, 196, 197, 212, 217, 219, xiv
Thompson, Roy, 151
throw-irons, 26, 27
Tioga Pass, 5
tip bender, 48, 101
Titus, Frank, 203
Todd, Tommy, 70, 100, 102, 126
Trans-Sierra mail, 75
Truckee, 29, 177, 183-185, 193-197
Truckee Ski Club, 201
Tubman, Father John, 150 to 201
Tull, Pell, 67, 100
Tuttle, Fred, 83, 93
Twain, Mark, 77

U
Ulland, Sig, 202
Union House Hotel, 58, 99, 100, 119, 215, 217
U.S. Army troops, 4
U.S. Bicentennial Observance, 95
U. S. Forest Service, 16-17 (map)
Utah Territory, 5

V
Vaage, Jakob, 83, 222
Van Clief, Judge Peter, 121
Varney, Sheldon, 84, 94
Vaughn, Jerome A., 174
Viking Ski Club, 202
Virginia and Truckee R.R., 196
Volunteer Corps, 108-110

W
Walker, Hiram, 70, 100
Walker, Laura, 34
Walker, Webster, 34
Walker, William, 143
Wallace, Ben, 105
Wallace, Hi, 119
Ward, Hamilton, 11, 26, 219
Ward, John, 119

Warner, Sen. John, 95
Wash, Robert, 61
Washoe Fever, 146
Watson, John 95, xv
Weaver, Mrs. Truman, xiii
Webb, Matthew, 111
Weir, William, xii
Wells Fargo History Room, xiv
Wentworth, 219
Wertheimer, Mert, 222
West, Major Fraser, 93
Wheeler, Sam, 92
whipsaws, 26, 54
Whiskey Diggings, 15, 49, 51, 219, 221
White, Bill, 99
Whiting, Fenton B., 33, 113, 120, 123
Whitlock, Major James W., 23, 24
Wilkinham, Jim, 54
Willard, H. L., 109, 110
Willard, E.X., 109
William B. Berry Western American Ski Sport Museum, 38, 79, 112, 200, xiv
Williams family, 101-103, 219
Williams, Ed, 103
Williams, Johnnie, 103, 104
Willis, Rev. Patrick Henry, 169, 170
Wilsdorf, M., 52
Wilson, Bill, 120,122
Wilson, James St. Clair, 170
Wilson, J.E., 226
Winter, Robert, 52
Winter Olympics, 195, 202
Wolfe and Company, 52
Wolfe, Hans, 83, 93
Wolfe, Jacob, 52
Wolfe, John, 53
Wood, William, 118
Woodward, Absolam, 74
Woodward, Alice, 105, 134
Woodward, Art, 57, 100, 105, 134, 137
Woodward, Birdie, 105
Woodward, Florence, 105, 134
Woodward, Frank, 57, 105, 134, 137
Woodward, George, 133, 134
Woodward, John, 57, 100, 105, 134, 137
Woodward, Martha, 134
Wooster, Clarence M., 180

Y
Yard, Charley, 101
Yellowstone National Park, 4
Yuba Gap Snow-shoe Club, 221
Yuba Pass, 209
Yuba River, 2, 13, 18, 19, 140,

Z
Zacharia Granville Express Company, 32
Zchocko Technic Institute, 37
Zent, Ike 51, 52
Zimmerman, B.M. 85